DISCOVER BONES

EXPLORE THE SCIENCE OF SKELETONS

By Lesley Grant
Illustrations by Tina Holdcroft

A ROYAL ONTARIO MUSEUM BOOK

Addison-Wesley Publishing Company

Reading, Massachusetts   Menlo Park, California   New York
Don Mills, Ontario   Wokingham, England   Amsterdam   Bonn
Sydney   Singapore   Tokyo   Madrid   San Juan
Paris   Seoul   Milan   Mexico City   Taipei

Many of the designations used by manufacturers and sellers to distinguish their products are claimed as trademarks. Where those designations appear in this book and Addison-Wesley was aware of a trademark claim, the designations have been printed in initial capital letters (e.g., Plasticine).

Neither the Publisher nor the Author shall be liable for any damage which may be caused or sustained as a result of conducting any of the activities in this book without specifically following instructions, conducting the activities without proper supervision, or ignoring the cautions contained in the book.

Library of Congress Cataloging-in-Publication Data

Grant, Lesley.
    Discover bones : explore the science of skeletons / by Lesley Grant ; illustrations by Tina Holdcroft.
        p.   cm.
    "A Royal Ontario museum book."
    Includes index.
    Summary: A wide-ranging look at bones, covering their structure and function, as well as such topics as growth, animal skeletons, archaeology, and fortune telling, with related activities.
    ISBN 0-201-63237-3
    1. Bones — Juvenile literature. [1. Bones.   2. Skeleton.]
I. Holdcroft, Tina, ill.   II. Title.
QM101.G69   1992
596'.0471 — dc20   92-30117
                                                                                                                                            CIP
                                                                                                                                             AC

Text copyright © 1991 by The Royal Ontario Museum
Illustrations copyright © 1991 by Tina Holdcroft

All rights reserved. No part of this publication may be reproduced, stored in a retrieval system, or transmitted, in any form or by any means, electronic, mechanical, photocopying, recording, or otherwise, without the prior written permission of the publisher. Printed in the United States of America.

Originally published in Canada by Kids Can Press, Ltd., of Toronto, Ontario.

Edited by Elizabeth MacLeod
Designed by N. R. Jackson
Set in 12-point Usherwood by Techni Process Limited

1 2 3 4 5 6 7 8 9-AL-95949392
First printing, September 1992

Addison-Wesley books are available at special discounts for bulk purchases by schools, corporations, and other organizations. For more information, please contact:

Special Markets Department
Addison-Wesley Publishing Company
Reading, MA 01867
(617) 944-3700 x 2431

Text stock contains over 50% recycled paper

Contents

The bare bones	**8**
Feel your bones	10
Journey to the center of the bone	12
Rubber bones?	14
Feed your bones	16
Lost in space	18
So how ya growing?	20
How tall will you be?	22
Spine tingling	24
Joints–where bones meet	26
Get moving	28
The hand that changed the world	30
Oh my achin' feet	32
Bony skyscraper	34
Them's the breaks	36
Those amazing flying machines	40
Who needs bones anyway?	44
A bone-hard puzzle	46
Tales bones tell	**48**
Digging up our past	50
Time travellers	52
Dogs and their bones	54
Get digging!	56
Bone detectives	60
Finding fossils	62
Read any good bones lately?	64
Only your archaeologist knows for sure	68
Special bones for special jobs	70
You look just like...your bones	72
How we use bones	**74**
Spirits in bones	76
Bones are beautiful	78
Cleaning bones	80
Skeletal jewellery	82
Medical miracles, sticky stuff and more	84
Foretelling the future	86
Bone mysteries	88
Old games with a new look	90
Glossary	92
Activities and experiments	92
Index	93
Answers	96

ACKNOWLEDGEMENTS

There is a hidden world at the ROM — a world that the visiting public rarely sees. Behind the scenes and deep in the bowels of this building are many people — curators, assistant curators, technicians, educators who take care of business. This means attending to the vast collections of artifacts, researching, developing educational programs and displays for the public and many other tasks. I'd like to thank Hugh Porter, who provided the "passports" and "road maps" necessary to navigate this inner world. My special thanks to people I visited repeatedly for advice and clarification: Peta Daniels, Kevin Seymour, Dr. Howard Savage, Ken Lister, Arni Brownstone, Susan Woodward, Janet Waddington, Paul Toth, Jim Dick and Marilyn Jenkins. Thanks also to Gordon Edmund, Roberta Shaw, Elaine Rousseau, Debra Reierson, Ross MacCulloch, Barbara Ibronyi, Kat Mototsune, Lynne Kurylo, Dr. James Hsü, Patty Proctor, Dr. Jeanne Cannizzo, Dr. Ian McGregor, Dr. Nick Millett, Dr. Glenn Wiggins, Dr. Edwin Crossman, Dr. Kathryn Coates, Dr. Judith Eger, Anne MacLaughlin, Dr. Peter Storck, Mark Peck, Philip Mozel, Dr. Angela Sheng and Dr. Chris McGowan.

My thanks as well to people outside the ROM for their contributions: Dr. Wayne Marshall; Dr. Martin Tammemagi; Jeanie Nishimura; Trudy Duff; Peter Glover; Kelly Robazza, D.C.; Bette Clark; David Sills; Hartley Miltchin, D.P.M; Colleen Zilio; Dr. Alastair Summerlee; Dr. Ken McCuaig and Dr. Rufus Crusher.

At times, I couldn't think of a nicer way to make a living than writing a book for children on such a fascinating topic. But at other times I began to understand why authors feel compelled to thank their mothers, great-grandfathers and pet iguanas for getting them through the rough times. For those times, I'd like to thank my children, Emily (who provided many of the ideas in this book) and Norelle (who tested the ideas enthusiastically), my husband, Peter, and my mother, Mary Lail. I'd also like to thank my editors, Liz MacLeod and Val Wyatt, for their encouragement and cheery inquiries, such as asking me if I was "bone weary" yet.

To my children, Emily and Norelle,
two bold and gentle questors…

The bare bones... about your bones, animal bones and no bones at all...

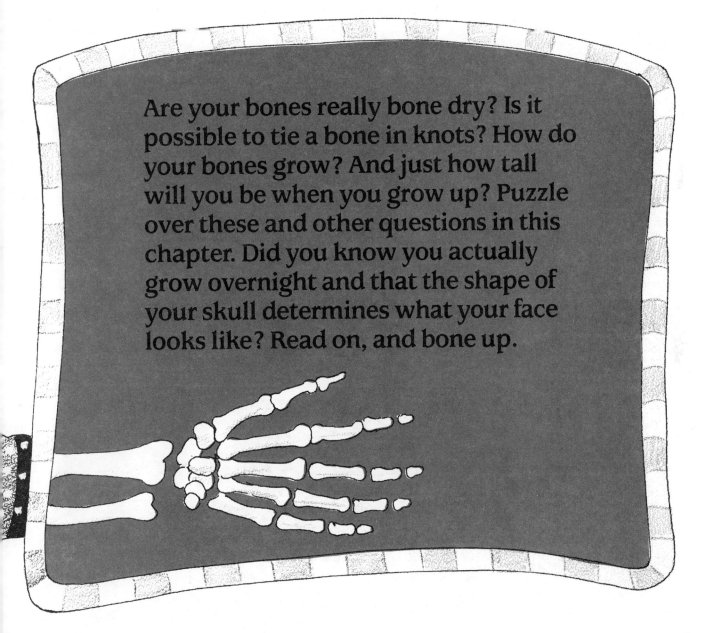

Are your bones really bone dry? Is it possible to tie a bone in knots? How do your bones grow? And just how tall will you be when you grow up? Puzzle over these and other questions in this chapter. Did you know you actually grow overnight and that the shape of your skull determines what your face looks like? Read on, and bone up.

Feel your bones

YOU CAN'T REALLY see your bones but you can feel a lot of them. Try feeling the bones in your hands. How many do you think you have? Can you find your tail bone? (Think about the last time you fell skating or tobogganed down a big hill. Now do you remember where your tail bone is?) What does it feel like? How many ribs can you count? There's no doubt about it—you're bony.

But billions of years ago, all creatures were soft and slippery and there wasn't a bone in sight! Scientists can't tell us why things got so bony, but they can tell us that over thousands of years new characteristics, such as a little more fur or a slightly different colour, will pop up from time to time. About 450 million years ago, a bit of bone just happened to appear in the skin of an animal. This slightly bony skin helped the creature survive long enough to reproduce and have slightly bony babies! Many animals today still have this suit of armour. For instance, crocodiles have bony plates just beneath their skin, and a tortoise's shell is actually made of bone.

Scientists still don't know how our bones moved inside our bodies. The fossil record shows that about 530 million years ago, some fish-like creatures with a spinal cord appeared. Gradually, vertebrae (backbones) developed to protect this cord.

One thing we do know is that an internal skeleton is handy. It lets you move about more freely than snails or worms can, for instance, and because it supports and holds up the soft parts of your body, it lets you grow bigger than most creatures without skeletons.

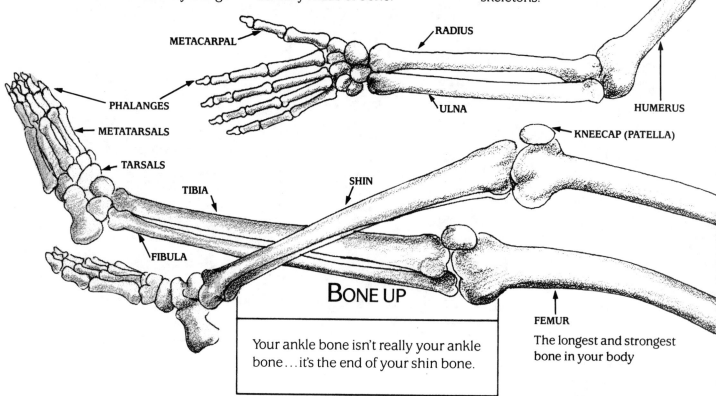

FEMUR
The longest and strongest bone in your body

Bone up

Your ankle bone isn't really your ankle bone... it's the end of your shin bone.

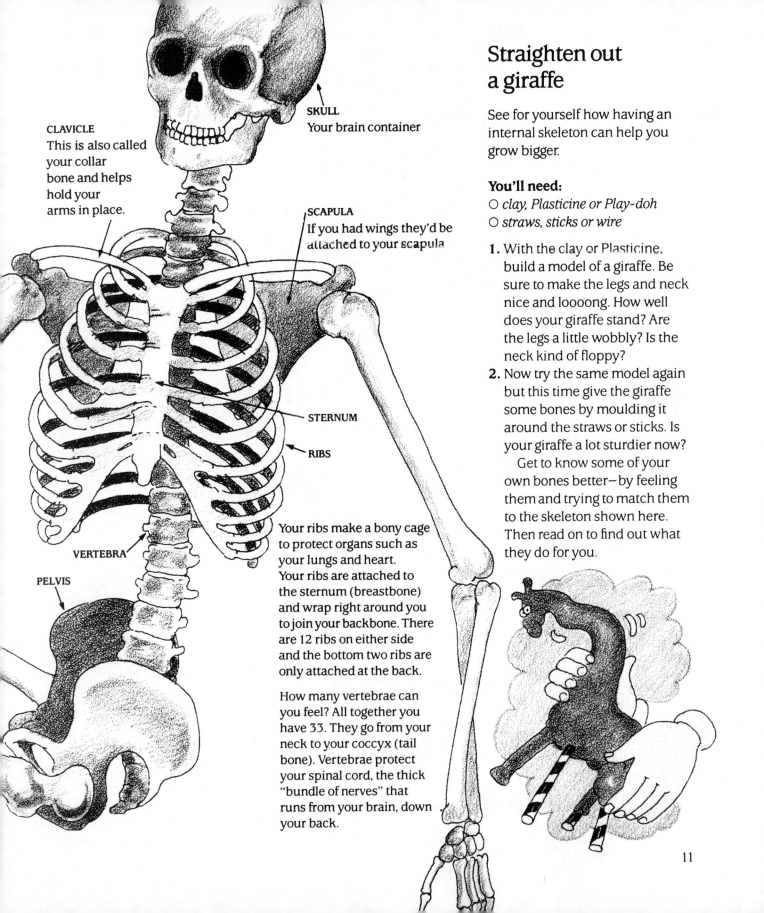

CLAVICLE
This is also called your collar bone and helps hold your arms in place.

SKULL
Your brain container

SCAPULA
If you had wings they'd be attached to your scapula

STERNUM

RIBS

VERTEBRA

PELVIS

Your ribs make a bony cage to protect organs such as your lungs and heart. Your ribs are attached to the sternum (breastbone) and wrap right around you to join your backbone. There are 12 ribs on either side and the bottom two ribs are only attached at the back.

How many vertebrae can you feel? All together you have 33. They go from your neck to your coccyx (tail bone). Vertebrae protect your spinal cord, the thick "bundle of nerves" that runs from your brain, down your back.

Straighten out a giraffe

See for yourself how having an internal skeleton can help you grow bigger.

You'll need:
○ clay, Plasticine or Play-doh
○ straws, sticks or wire

1. With the clay or Plasticine, build a model of a giraffe. Be sure to make the legs and neck nice and loooong. How well does your giraffe stand? Are the legs a little wobbly? Is the neck kind of floppy?
2. Now try the same model again but this time give the giraffe some bones by moulding it around the straws or sticks. Is your giraffe a lot sturdier now?

Get to know some of your own bones better—by feeling them and trying to match them to the skeleton shown here. Then read on to find out what they do for you.

JOURNEY TO THE CENTER OF THE BONE

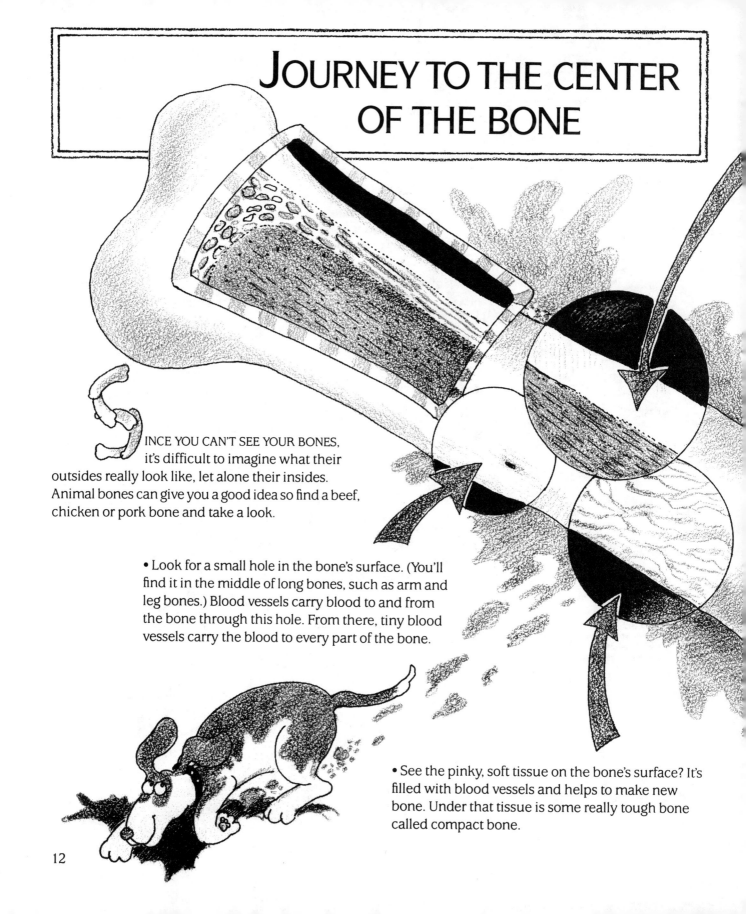

SINCE YOU CAN'T SEE YOUR BONES, it's difficult to imagine what their outsides really look like, let alone their insides. Animal bones can give you a good idea so find a beef, chicken or pork bone and take a look.

• Look for a small hole in the bone's surface. (You'll find it in the middle of long bones, such as arm and leg bones.) Blood vessels carry blood to and from the bone through this hole. From there, tiny blood vessels carry the blood to every part of the bone.

• See the pinky, soft tissue on the bone's surface? It's filled with blood vessels and helps to make new bone. Under that tissue is some really tough bone called compact bone.

- The middle of some bones is filled with a soft substance called marrow. It's easiest to see the marrow in a chicken leg bone or a rib or vertebrae.

There are two kinds of marrow: red and yellow. The red marrow is found mainly near the ends of your long bones (such as those in your arms and legs) and in your ribs, vertebrae, pelvis and breastbone. Its job is to make blood cells, about 200 billion each day. The yellow marrow is found in the shafts of long bones and its job is to store fat.

- Under the compact bone is bone that looks like honeycomb, called spongy bone. You can see this best if you crack open a chop or rib bone. (Have an adult help you.) Spongy bone strengthens your bones but because it's full of spaces, it doesn't add a lot of extra weight. If your bones were made only of compact bone, you'd be very heavy!

Native people have known for a long time that bones contain marrow and fat. They would break bones open and extract the marrow with a hooked tool. Since marrow has a lot of blood cells, it's very nutritious. Or, natives used the fat in the bones. They broke the bones, then boiled them. When the fat or grease rose to the surface of the water, they skimmed it off and mixed it with meat or used it as butter.

Bone brush

The First Nations of the Plains used the spongy part at the end of bones for paint brushes since it could absorb and hold paint. What's it like painting with a bone? Try it and find out.

You'll need:
○ *a knife or hacksaw (get an adult's help)*
○ *a variety of bones, such as long bones (leg or arm bones), round bones, etc.*
○ *thick paint*
○ *paper*

1. Cut the end off a long bone to expose the spongy-looking bone inside.
2. Dip the long bone in paint and use as a paint brush.
3. Dip the other bones you have in the paint and use them to make designs on the paper.

RUBBER BONES?

WIGGLE THE TIP OF YOUR NOSE AND ears. Do they feel rubbery? That's because they're made of cartilage, a tough, rubbery substance. When you were a baby, your bones were soft because they were mostly cartilage. As you grew, this cartilage gradually hardened as the minerals calcium and phosphorus were added. Now your bones are made of collagen (the main ingredient in cartilage) and calcium phosphate (a combination of calcium and phosphorus). Together, calcium phosphate and collagen make your bones extremely strong.

What do you think happens to bone if you remove either the calcium phosphate or the collagen? Try these experiments and see for yourself.

Bend a bone

You'll need:

○ a chicken leg or rib
○ a cloth
○ vinegar
○ a small jar with lid

1. Scrape the meat off the bone. Wash and dry it with the cloth.
2. Now try to bend the bone. Can you? Does it flex at all?
3. Pour the vinegar into the jar, add the bone, and put the lid on. Leave the bone for two weeks, then remove it and rinse it with water.

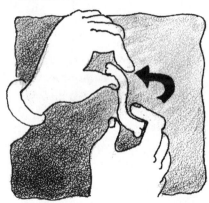

4. Now try bending the bone. How far can you flex it? Can you tie a knot in it? (Put your bone back in the vinegar for a week if it's still too stiff.)

What happened to the bone? The calcium in it dissolved in the vinegar, leaving behind a bone made of soft, rubbery collagen.

If you don't have a bone, try this experiment with an egg. Place the egg in the vinegar for about a week. The egg shell will become very rubbery as the calcium leaches out.

Bake a bone

In this experiment, you'll make a bone brittle by removing some of the collagen.

You'll need:

○ a pork, beef or chicken bone (you can get it at a butcher's or in the meat department of your grocery store)
○ a strong plastic bag
○ safety glasses
○ a hammer (get an adult's permission to use it)
○ a scale

1. Bake the bone in a 175°C (350°F) oven for three hours. (If you do this when something else is cooking, you can save energy.)
2. Take the bone out of the oven and let it cool.
3. Place the bone in a strong plastic bag and put on some safety glasses. Then hit the bone with a hammer a few times. The bone may splinter so it's important to contain the pieces in the bag and protect your eyes. What happened to the bone? Baking it destroyed the bone's collagen, leaving the bone very brittle.

If you weigh the bone before and after you bake it, you'll find it weighs less afterwards. Why? Bones are 25-50% water! The water evaporates as the bone is baked.

BONE UP

How are your bones and the head of a match alike? They both contain phosphorus. In fact, the bones in your body have enough phosphorus in them to make 2000 match heads!

FEED YOUR BONES

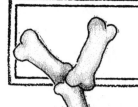

YOUR BONES contain over 1 kg (2 pounds) of calcium, or about the same weight as four big apples. Sounds like calcium is pretty important to your body, doesn't it? The work of your heart, muscles, nerves and blood depends on calcium and so does the strength of your bones. If you did the experiment on the previous page, you discovered that bones become weak and rubbery without calcium.

Before we understood the importance of calcium and vitamin D in a healthy diet, many children and adults developed rickets, a disease that makes bones soft. The bones become too weak to support the person's weight and they actually bend and bow out under the strain. Bow legs is one sign of rickets.

Rickets still occurs today but it's unusual in countries where kids eat and drink lots of milk, cheese and other high calcium foods. Because your bones are growing rapidly, you need more calcium, phosphorus and vitamin D than adults do. Have you ever noticed that milk containers say "Vitamin D added"? It's there to help your body absorb the calcium. Sunlight is a great source of vitamin D but the short winter days in countries such as Canada can prevent you from getting all the vitamin D you need.

Calcium cooking

Your bones are like a store. But this store is stocked full of calcium, not cans of spaghetti. If your body needs a little extra calcium, such as when you're going through a growth spurt or your body is trying to mend a broken bone, it can dip into those supplies. You're probably getting all the calcium you need, but why not try a few of these recipes and stock the shelves with some extra calcium!

CRUNCHY CALCIUM COOKIES
These tasty no-bake treats are fun to make.

You'll need:

○ 125 mL (1/2 cup) crunchy peanut butter
○ 125 mL (1/2 cup) honey
○ 125 mL (1/2 cup) powdered milk
○ a bowl
○ a fork
○ a plate

1. With the fork, mix the peanut butter and honey together in the bowl.
2. Add the powdered milk and mix well. (You might have to squish the mixture a bit with your hands to make it stick together.)
3. Form the mixture into bite-size balls and place on a plate.
4. Put them in the freezer for two hours.
5. Eat!

Makes about 24 bite-size balls.

MOLASSES MILKSHAKE

Molasses and yogurt are great sources of calcium.

You'll need:

○ 250 mL (1 cup) milk
○ 50 mL (1/4 cup) yogurt
○ 1/2-1 banana or equivalent of your favourite fruit
○ 30 mL (2 tbsp) or more of molasses
○ a blender

Mix ingredients together in the blender.
Serves two.

SALMON MELT SANDWICH

There's lots of calcium in salmon and cheese.

You'll need:

○ 1 can of salmon
○ 1/2 stalk of celery, chopped
○ 15 mL (1 tbsp) mayonnaise
○ bread or bun
○ grated cheese
○ a bowl

1. Drain the salmon and empty it into a bowl. (Like all bones, the tiny bones in salmon are very high in calcium. If you like, mash them up well and mix with the salmon.)
2. Add the celery and mix in the mayonnaise.
3. Spread onto the bread or bun and sprinkle with grated cheese.
4. Warm in a microwave or grill in the oven. (Get help from an adult if you need it.)

BONE UP

You can always eat kelp (seaweed) for a quick fix of calcium. It's one of the richest sources of calcium available. About 250 mL (1 cup) of kelp contains 2405 mg (1 ounce) of calcium. (A glass of milk contains about only about one-third as much.) You need 500-700 mg of calcium every day. Here are some good sources of calcium you can depend on.

Food	Calcium	Food	Calcium
15 mL (1 tbsp) cheddar cheese	53 mg	125 mL (1/2 cup) skim-milk powder	828 mg
15 mL (1 tbsp) molasses (blackstrap)	137 mg	250 mL (1 cup) salmon, canned	888 mg
250 mL (1 cup) green leafy veggies	200 mg	250 mL (1 cup) kelp	2405 mg
250 mL (1 cup) skim milk	298 mg		

LOST IN SPACE

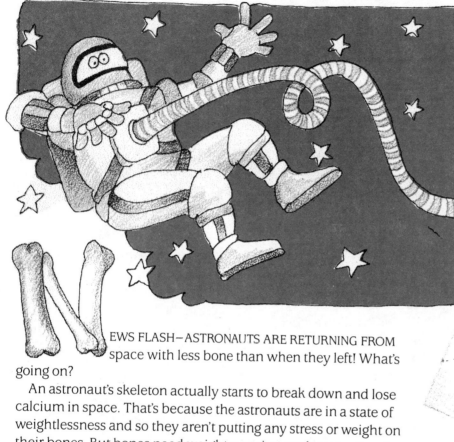

NEWS FLASH—ASTRONAUTS ARE RETURNING FROM space with less bone than when they left! What's going on?

An astronaut's skeleton actually starts to break down and lose calcium in space. That's because the astronauts are in a state of weightlessness and so they aren't putting any stress or weight on their bones. But bones need weight pressing on them to stimulate new growth. They need the push and pull of gravity to stay strong and healthy.

Scientists worry that after a very long voyage, an astronaut's skeleton may have lost so much calcium and be so weak that it could shatter upon re-entry into the earth's atmosphere. You actually get more of a workout walking from your bedroom to the kitchen than an astronaut gets all day! So the doctors have developed exercise programs to put stress on the astronauts' bones. High above the earth, astronauts work out at least two to three hours every day by strapping themselves into a special harness that holds them down and provides a force to work against, the same way gravity does on earth. They also anchor themselves to treadmills for running and pull on special springs to exercise their arms.

Back here on earth, the bones of bedridden people become thinner and weaker for the same reasons astronauts' bones do—they aren't being used. The bones actually become smaller as the calcium in them breaks down and is carried away in the bloodstream.

So in or out of this world, exercise is the key to strong bones. Did you know that athletes' bones are much heavier and stronger than other people's? You are probably pretty active and get all the exercise you need, but sometimes during long winter months you might get a bit lazy. If that's your problem... get busy and exercise your bones!

BONE UP

Older people, especially women, often develop weak, porous bones. Doctors called this condition osteoporosis. The bone contains less calcium so it gets weaker and more brittle and breaks more easily. Experts still aren't sure how to prevent osteoporosis but they think eating a calcium-rich diet, exercising and not smoking all help.

SO HOW YA GROWING?

YOU'VE PROBABLY noticed that 12-year-old girls often tower over 12-year-old boys. But by the time boys turn 18, they're usually taller than the girls. The answer to this puzzle is at the ends of your bones.

A lot of the growth in your bones happens at the ends. These ends, or epiphyses as scientists call them, are made of soft cartilage and there is lots of space that allows growth to take place. When you are finished growing, the ends of the bones harden and the space fills in. Girls finish their growing first, when they're between 12 and 14, and the ends of their bones harden years before boys' bones do. What this means is that boys have more time to grow and so often grow to be taller.

Speaking of growing, are you feeling a little stiffer than usual? Maybe you're in the middle of a growth spurt. During a spurt, your tendons and other soft tissues don't grow as quickly as your fast growing bones. So your joints may be a little tight and you may not be as flexible. This could be one reason why kids get growing pains in their joints—from those fast bones!

Some bones grow so fast that it takes years for other bones to catch up. During puberty, your foot bones grow faster than any other bones and reach their final size long before the rest of you does.

Your collar bone and breastbone fuse when you're around 25 years old.

The epiphyses at your elbows start to fuse when you're 12 to 13 years old and finish when you're around 16.

The ends of your hip and ankle bones begin fusing when you're about 14 and are finished by the time you're 19 years old.

At your knees, the epiphyses start fusing when you're about 16 years old and finish by the time you're 22.

Grow overnight

Here's a 100% guarantee that you will grow tonight. Actually you grow every night... and you shrink the next day! What's going on? Does the sun make you shrink? Does the moon make you grow?

It all happens in your spine or backbone. Run your hand down the middle of your back. Those bumps you feel are little bones called vertebrae. (Because of them, scientists classify you as a vertebrate, an animal with a backbone.) In between each vertebra is a soft pillow of cartilage called a disc. Those pillows are your spine's shock absorbers.

When you sit, stand, walk or run, your discs are squeezed together between the vertebrae. The discs prevent the vertebrae from rubbing against each other so that you can bend and stretch easily and painlessly. Your running shoes have shock absorbers that are a bit like your discs. When you walk, the heel of the shoe compresses to cushion your foot. When you aren't putting pressure on your shoe, the heel expands again.

That's what happens to the discs in your back, too. After a hard day of playing, your discs have been compressed, and you may be 6 to 12 mm (¼ to ½ inch) shorter than when you woke up. Here's how you can prove it.

Sleep and grow

You'll need:

○ a wall
○ a friend to help you measure accurately
○ a ruler

1. Remove your shoes. Stand with your back against the wall and have your friend place a ruler on your head. Mark where the ruler meets the wall. Do this: (a) before you go to sleep, (b) in the morning, (c) in the afternoon.
2. Look at your measurements. When were you tallest? Was there a difference between morning and night? What about morning and afternoon? What are your discs doing during the day and at night-time?

BONE UP

When you were a baby, most of your bones were in separate pieces, which gave you lots of space to grow. As you grew, more and more of these pieces fused together. A baby's skull, for example, has six gaps between the bones that allow the baby's head to grow, as well as compress when it is being born. Those gaps are the soft spots you can feel on a baby's head. But by the time the baby is two, the bones have grown together, leaving a pattern of squiggly joints where they fused. Take a look at the illustration on the left to find out when you'll start to fuse together!

How tall will you be?

The tallest man, Robert Wadlow, was 272 cm (8 feet, 11 inches) tall or about twice as tall as you are! The shortest person that we know of was Pauline Musters of Holland. When she was fully grown, she would have stood only about as high as your knee.

YOUR HEIGHT depends on how much your bones grow—they lead the way. You inherit a tendency to be tall or short from your parents but what you feed your bones is important, too. Your skin and muscles need lots of protein to grow, and your bones need lots of calcium. Your bones lead the way and provide the structure for the rest of you.

Sometimes people grow to be exceptionally tall and sometimes they don't grow very much at all. One reason people don't grow is that their bodies don't produce enough growth hormone and so their bones grow very slowly. A child who doesn't have enough growth hormone sometimes receives injections of it to help him grow. If a person's body produces too much growth hormone, her bones are stimulated to grow longer very quickly and she becomes exceptionally tall.

HOW BIG WILL YOU BE?

Here's one way to estimate how tall you'll end up and how much you'll weigh. It's only a guess so don't get discouraged if you don't like the answer. Also, this simple formula can't take into account whether you are maturing slowly or quickly and that makes a difference.

If you are a boy, find out:
- your height at age two
- your weight at age two

Now multiply your height at age two by 2. Multiply your weight at age two by 5.

If you are a girl, find out:
- your height at one and one-half years
- your weight at one and one-half years

Multiply your height at age one and one-half by 2.
Multiply your weight at age one and one-half by 5.

If your parents don't have a record of your height and weight, your doctor probably does.

One family calculated the final height for their four young boys and discovered that all their children would be at least 2 m (6 feet, 4 inches) tall. They lengthened the boys' beds to 2.4 m (7 feet, 9 inches) just to be safe.

KEEPING TRACK

Your parents may have a record of your growth. If so, you could plot these figures on a graph to get a picture of just how fast or slow you are growing at different times.

Even if you don't have these numbers, it's not too late to begin. Start keeping track of your growth now. Once or twice each year, measure yourself and record your height on a stick or chart. Then if you're feeling especially short one day, you can check your record and see how fast you are growing.

BONE UP

Your bones make up roughly one-fifth of your body weight. Some people claim that they're not really overweight, it's just that their bones are extra heavy! They might be right. Bones vary from person to person and some people do have bones that are thicker and heavier than average.

SPINE TINGLING

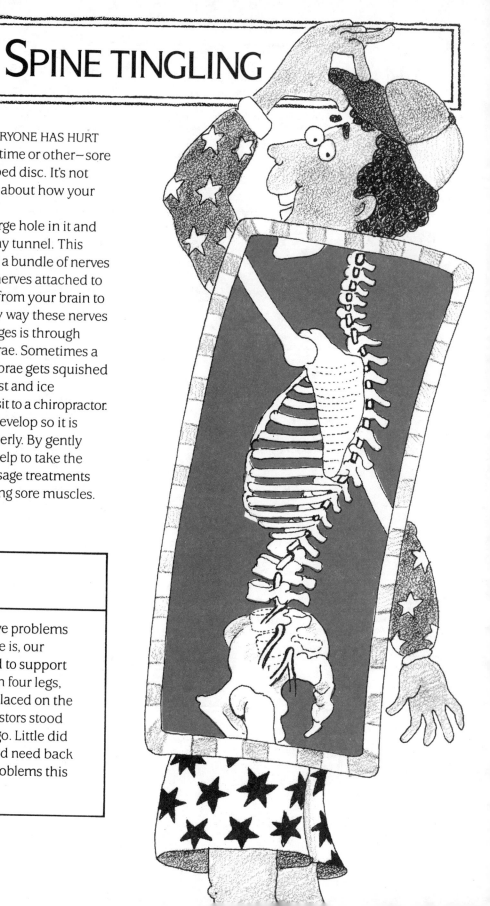

IT SEEMS AS IF EVERYONE HAS HURT her back at some time or other—sore muscles, a pinched nerve, a slipped disc. It's not really surprising when you think about how your back is put together.

Each of your vertebrae has a large hole in it and these holes line up to make a bony tunnel. This tunnel provides a safe passage for a bundle of nerves called your spinal cord. Smaller nerves attached to your spinal cord carry messages from your brain to the rest of your body. But the only way these nerves can get out to deliver their messages is through tiny gaps in between your vertebrae. Sometimes a cushiony disc between two vertebrae gets squished and presses on a nerve. Ouch! Rest and ice treatments will help. So might a visit to a chiropractor. A chiropractor helps your spine develop so it is strong, flexible and lined up properly. By gently moving your vertebrae, she can help to take the pressure off pinched nerves. Massage treatments can help your back, too, by relaxing sore muscles.

BONE UP

Why do so many people have problems with their backs? The trouble is, our back was originally designed to support our ancestors who walked on four legs, not two. New pressure was placed on the lower back when these ancestors stood up many millions of years ago. Little did they know that someday we'd need back doctors to take care of the problems this new posture would create!

Treat your back right

There are things you can do to take care of the muscles, nerves, vertebrae and discs in your back. Choose the picture you think would be best for your back. Think about the positions the dog is in. Is his back supported?
Answers on page 96.

STRAIGHT UP!

How good is your posture? Find out with this test.

1. Stand with your back against a wall.
2. Try to touch the wall with your shoulder blades, bottom and heels.
3. While standing in this position, answer the following questions.
— Can you fit your hand in the gap between the wall and your lower back and wiggle it about?
— Do your shoulders curl forward, away from the wall?
— Is there a big space between the wall and your neck?

If you answered "yes" to any of these questions, then it's time to work on your posture by strengthening your back and stomach muscles. Here are some exercises you can do that will help.

25

Joints...Where Bones Meet

WITH MORE THAN 200 BONES packed into your body, your bones are going to run into each other a lot. But instead of colliding, there are 230 places called joints where your bones meet. Knees, elbows, wrists, ankles—these are just some of the joints in your body and without them you'd be stiff as a steel rod. Stand very straight and rigid. Now...can you move without bending any joints? Try holding this book without bending your joints. Without joints you'd be like the Tin Man—stuck!

Different parts of you can bend, twist, open, close, stretch and much more, so you need different types of joints for all these various movements. See if you can find the three types of joints shown here in your body. You'll have to swing and bend your arms and legs, move your head, hands, feet, and bend your back, to discover them all.

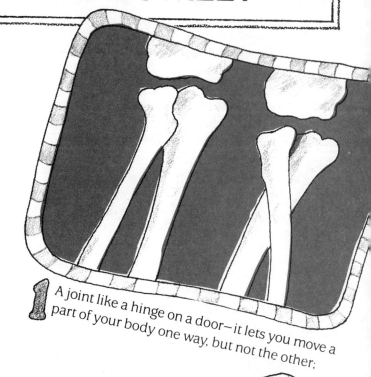

1. A joint like a hinge on a door—it lets you move a part of your body one way, but not the other;

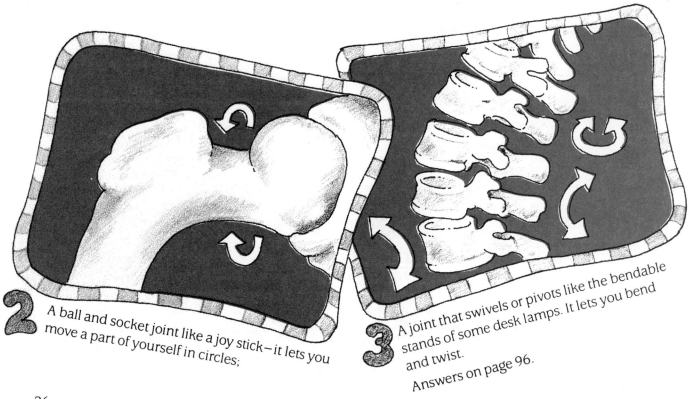

2. A ball and socket joint like a joy stick—it lets you move a part of yourself in circles;

3. A joint that swivels or pivots like the bendable stands of some desk lamps. It lets you bend and twist.

Answers on page 96.

You have 22 bones in your skull but only one movable pair of joints. Can you find it? Put your fingers on your chin and open and close your mouth. Now can you find it? The rest of your skull bones are held together by immovable joints called sutures. They look like wiggly lines on your skull and allow growth to take place at the edges of the bone.

Your joints need to be kept oiled but unlike the Tin Man, your joints already have some lubrication built into them. Your body's hinges are covered with a very smooth layer of cartilage. A lubricating liquid called synovial fluid is produced right in the joint, allowing the bones to slip freely over one another. And the ends of your bones are covered with a smooth elastic tissue called cartilage that also prevents bones from rubbing each other the wrong way! But if the cartilage wears away and there isn't enough fluid to lubricate the joint, the rough ends of the bones rub together, and cause a condition called osteo-arthritis.

Animals suffer from arthritis, too, now and long ago. For instance, scientists examining the backbone of a 3500-year-old dog noticed areas where the bones had worn away. They guessed that this was a sled dog's backbone and the heavy loads it pulled had caused arthritis. From examining dinosaur bones, archaeologists (experts who study people and animals from the past) think dinosaurs may have had arthritis, too!

THE SHUTTLE'S ARM

When you rotate your wrist, you can feel the bones in your forearm, the ulna and the radius, crossing over each other, letting you twist your hand around. The mechanical arm on the U.S. space shuttle, the Canadarm, has joints that twist and swivel in the same way. It also has wrist, shoulder and elbow joints, just like you. However, unlike you, the Canadarm can pick up loads in space as massive as 30 000 kg (66 000 lb.), about the same as two fully loaded buses!

BONE UP

Scientists aren't sure what makes all that racket when you "crack" your knuckles, or your joints pop. Some experts think the gas in your joints expands and forms bubbles, which pop. Others think the cracking occurs as your tendons and ligaments slip over each other.

Get Moving

IT MAY BE A LONG time since you last did the "bunny hop," but you probably still remember that you weren't very good at it. You couldn't move nearly as fast as a rabbit or hare and you couldn't do it for very long.

Next time you're at a museum, you can find out why a hare is a natural for hopping and you're not. Find a hare skeleton and take a close look at its hind leg bones. Now, compare them to the leg bones of a human skeleton. (Can't find a human skeleton? A bear's skeleton is very similar.) You will see that the hare is always ready to leap even when it's resting or standing still. Its huge hind leg bones are like springs, coiled and ready under its body. But our leg bones are already stretched right out.

So even though we try to bunny hop, we just can't seem to do it. Not only are our legs all wrong, our pelvis isn't right either! The hare's pelvis is very long and carries the force of the leap to the rest of the body. Because we walk upright, our pelvis is shorter and rounder.

And you'll never see a turtle hopping like a frog either. There are a number of bony reasons for that. Take a look at the long leg bones of the frog. When the frog is sitting still, those bones are folded on top of each other like an accordion. When the frog jumps, the bones unfold completely, propelling the frog skyward. The frog also has a very short spine that can handle the stress of all this leaping about. But the turtle has one really big problem to overcome before it can get moving: its bony shell. Somehow it has to get its little legs far enough outside its shell to walk. It solves this problem by sticking its upper leg bones straight out first, then dropping the lower leg bones down to the ground.

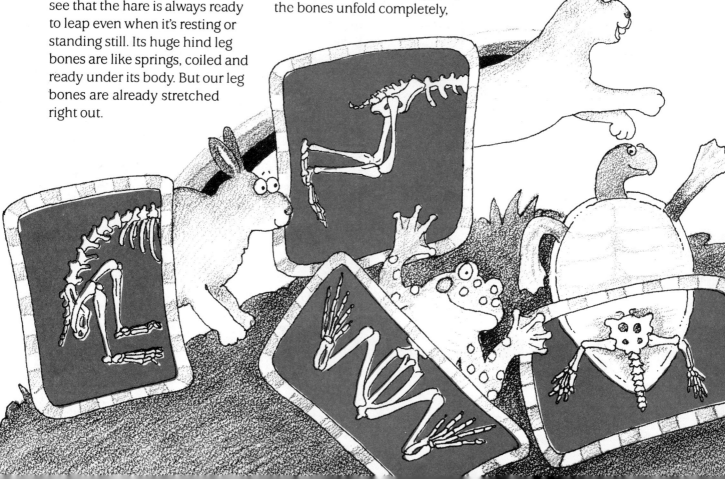

All animal species have different bony structures because they have all developed different methods of survival—methods for getting food and avoiding being someone's food. The turtle hides in its bony shell, the hare leaps away, and we stand...thinking inside our bony brain container.

It takes more than just bones to get you moving. You need muscles to pull the bones. And since muscles can pull in only one direction, bones are usually attached to pairs of muscles. One muscle pulls the bone in one direction, and the other muscle pulls it in the opposite direction.

Here are some body benders you can do to see just how your bones move.

1. Place your hand on your bicep, the muscle in your upper arm. Now bend and straighten your elbow. Can you feel the bicep move?

2. Put one hand on the back of your thigh and the other hand just above your knee. Now bend and straighten your knee and see if you can feel which muscle is doing the pulling when you bend and when you straighten your leg.

Tap your toes

Take your socks and shoes off and move your foot up and down as far as it will go. See those long thin lines just under your skin? Those are your tendons. They connect your bones to your muscles and without them your bones couldn't move. Can you see the tendons of your hands when you wiggle your fingers? Now feel the two large ropy tendons at the back of your knee.

Tendon tricks

To find out more about how tendons work, try this experiment.

You'll need:

○ a knife (get an adult's permission)
○ a chicken foot (ask at a butcher's) or chicken wing
○ tweezers

1. Cut away the skin around the leg of the chicken so that you can see the tendons. They look like thin white cords.
2. With your tweezers, grasp a tendon and give it a pull.
3. Pull each tendon separately and watch what happens.

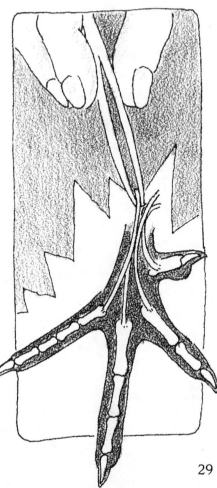

THE HAND THAT CHANGED THE WORLD

TOUCH EACH OF YOUR FINGERTIPS with your thumb. Easy, right? Well, maybe for you but you've just done something that only humans, apes and some monkeys can do. A unique joint—the saddle joint at the base of your thumb—allows you to make that special pincer grip and without the saddle joint, it's likely that humans would have evolved differently. We wouldn't have been the builders and inventors that we are because we wouldn't have been able to hold tools.

Artificial hands are designed to mimic this all-important pincer grip. This artificial hand has two fingers and a thumb that can be pinched together for grasping things. Find out how important the saddle joint is by trying the experiment on the opposite page.

Life without a saddle joint

You'll need:

○ *sticky tape*
○ *your hand*
○ *a banana, ball, spoon or pencil*

1. Tape each thumb to the palm of each hand.
2. Try to peel a banana, throw a ball, pick up a spoon or write the word "thumb."
3. Release your thumbs and tape your ring finger and baby finger to your palm. Now try the same activities again. Which way was easier?

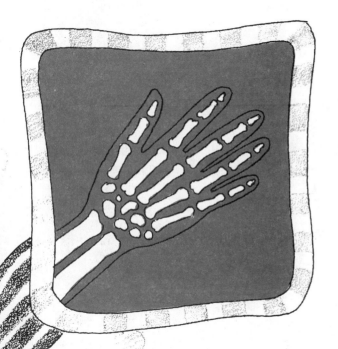

Handy muscles

Your hands are jam-packed with muscles, tendons and bones. In fact, there just isn't room for the 38 muscles needed to move the bones in your hands, so about half of them are found in your forearm. Put your hand on your forearm, make a fist and feel those muscles moving. If you look at the back of your hand, you'll see your tendons, those straight, raised lines running from your knuckles to your wrist and beyond. They connect your finger bones to the muscles of your forearm. Even with all these muscles, some of your fingers still have to share the same muscle.

See if you can figure out with a bit of finger flexing which finger bones have to share a muscle.

1. Make a fist.
2. Now extend your thumb without moving your other fingers.
3. Now try moving your index or pointing finger without moving any other fingers.
4. Try to extend your middle finger.
5. Now try to stick your ring finger out.
6. Move your little finger.

Did you discover that it's hard to move some fingers alone? Your ring finger and middle finger are moved by shared muscles, so they have to go most places together. Your index finger has its own muscles and so does your little finger. Your thumb is so important it has an entire set of muscles all to itself.

BONE UP

Porpoises with hands? Bats with fingers? It's true. Even though the flippers of the porpoise look like fins, there really are finger bones inside. Bats have finger bones, too. These bones provide a framework for the thin sheets of muscle and skin that make up its wing. The bat's "thumb" is a sharp hook that protrudes beyond the wing and is used for holding food, crawling and fighting.

Oh My Achin' Feet

How far do you think you will walk in your lifetime? The equivalent of walking across North America? Around the world? Twice around the world? Would you believe 160 000 km (99 500 miles) or roughly five times around the world? That's a lot of walking—about 15 000 steps a day. And when you run, the pressure on your feet is three to four times your body weight. So your feet have to be very strong.

In addition to containing one-quarter of all your bones—26 bones per foot—your feet have evolved an amazing system of arches to deal with all this pressure. The arch you probably know best is the one on the inside of your foot—the medial arch. There's another arch on the outside of your foot, and one more running across your foot, behind your toes. These arches give your feet the springy strength they need to absorb the weight of your body. Two thousand years ago, architects already knew how strong arches could be. The Romans used them in their buildings and bridges and many of these are still standing today! An arch is just the right shape to safely bear the stress of a heavy load.

Shoes then and now

The next time you are at the museum, have a look at the displays of artifacts from China. You may see examples of women's slippers that are too small for even a little child of today to wear. Some

Chinese women had extremely small feet because when they were young girls, their feet were tightly bound with cloth. As you can see in the X-ray below, the bones in the front part of

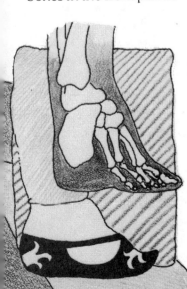

the foot and the heel were pushed together until eventually they stayed in that position. A woman whose feet had been bound like this couldn't walk normally. She hobbled along, often in pain. But her small feet were a sign that she was rich and didn't have to work.

Even today we punish our feet, squishing them into shoes that are too small, too pointy, or with heels as high as 15 cm (6 inches) or more. Bones get scrunched and crunched. So free your feet. Choose the shoe that's right for you. Here's how.

- New shoes should feel comfortable right away.
- If your feet are still growing, you need to leave some extra room at the front. Can you wiggle your toes? Can you press your thumb down in front of your big toe? You may be growing so fast that you'll be out of these shoes in three to six months, so leave room!

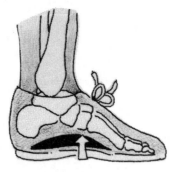

- Look for good arch support. If you weigh 40 kg (90 pounds), your feet are feeling the pressure of 160 kg (350 pounds) when you run, so arch support is important.

Some people have flat feet, which means they don't have an arch in the middle of their foot. If the "spring" ligament in the foot is weak, it can't hold up the bones of the arch, so the arch flattens. Are you flat footed? It's easy to find out.

SPRING LIGAMENT

Make a good impression

You'll need:
- baking soda
- a bath tub (make sure it's dry)

1. Pour a little baking soda in the bath tub and make footprints in it. Or after a bath, step onto a piece of paper while your foot is still wet.
2. Look at your footprints and compare them to the ones shown here.

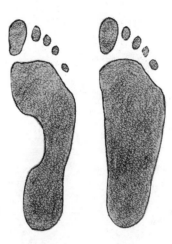

You might be flat footed (above, right) only when you are standing on your foot. A truly flat foot is one that is flat even when there is no weight on the foot at all.

Don't worry if your two-year-old sister looks flat footed. Most kids don't develop an arch until they are three or four. The trouble with flat feet is that too much weight is put on other parts of your body—like your knees. If you are flat footed, a special arch support can be made for your shoes.

33

BONY SKYSCRAPER

IT'S AMAZING HOW HIGH skyscrapers can stretch up into the sky. The building's framework, to which the walls, ceilings and floors are attached, has to be very strong. You are a bit like a skyscraper. Many of the same principles that architects and engineers use to design and build skyscrapers have been in use for millions of years in the human body.

Take a look at the Eiffel Tower and the bone shown next to it. The tower's girders are curved in the same way the head of the femur or thigh bone is. This curve means that no one spot bears any more weight than any other. And places where your bones have to bear extra weight like the ends of long bones, are reinforced with bony struts a lot like the Eiffel Tower's struts shown here. Bridges and airplane wings are also reinforced with struts.

The next time you pass a construction site, take a good look at the beams there. They have to be good and strong or they might break in the middle under the tremendous loads they carry. Engineers have developed ways of making beams twice as strong without making them twice as heavy. The shaft of your long bones is like a beam. There is a thicker layer of compact bone on the shaft to strengthen it.

The strength of any structure depends partly on the way the different pieces are held together. The places where these pieces connect are called joints, just like the ones in your body. What's the best method for joining these different pieces together? Some aircraft builders in the early 1900s thought that screws and nails would be best—after all, they worked for furniture. But the aircraft just ripped apart. You have to understand what you want the structure to do before you can choose the best joints for it. As you read on pages 26 and 27, your body has many different joints because each has a special job to do.

Strut some stuff

If you look closely, you'll see that the struts shown here are really triangles packed together. Take a look at the inside of a chicken wing or leg and you will see this triangular bracing, too. Architects have learned (and birds have always known) that triangles are stronger than squares. Find out for yourself.

BONE UP

Deep in your ear are the tiniest bones in your body: the hammer, anvil and stirrup bones. When sound funnels into your ear, the sound waves make your eardrum vibrate. The hammer bone is attached to your eardrum and so it vibrates too. These vibrations are passed along to the anvil bone and then the stirrup bone. The tiny stirrup bone is only 3 mm (1/6 inch) long but it has the important job of passing the sound vibrations on to the cochlea, a snail-shaped tube that carries the sound message to your brain. The bones of your skull also help you hear yourself speak since they vibrate to the sound of your voice.

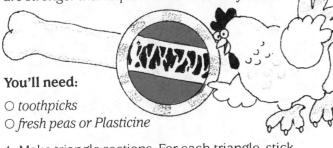

You'll need:

○ *toothpicks*
○ *fresh peas or Plasticine*

1. Make triangle sections: For each triangle, stick the end of a toothpick into a pea or a small ball of Plasticine. Put another pea or Plasticine ball at the other end of the toothpick and poke a second toothpick into it, as shown. Stick a third pea or ball at the end of that toothpick, then close your triangle by attaching a third toothpick to the peas or balls. Make a number of triangles like this.
2. Make square sections in the same way as described in step 1, but using four toothpicks and peas (balls) for each one.
3. Now build two different structures, one just with triangular sections, the other with square sections. Which is the easiest to build? Which is more stable?

THEM'S THE BREAKS

HAVE YOU EVER NOTICED HOW quickly your body goes to work to heal a scratch or cut? It reacts immediately by forming a clot to stop the bleeding. When you break a bone, the same thing happens. A clot forms at the break to temporarily hold the fracture together and stop the bleeding. Yes, your bones do bleed! But the good news is that even as you're being driven to the hospital, osteoclast cells—the bone cleaners—are busy dissolving the jagged edges of the bone, preparing the way for the osteoblast cells—the bone builders.

Osteoblasts work round the clock to produce new soft bone cells for you. Over the next two to three weeks, these cells will form a bridge of soft bone to connect the fractured ends. Then your body dips into its calcium stores to turn the soft bone into hard bone. Some bone fractures are very simple to repair but others are more difficult because the bone or skin is much more damaged.

Chances are you know someone (maybe yourself) who's broken a bone and had to wear a cast. But casts are a fairly recent way of treating breaks. Long ago people had to just let a broken bone heal by itself. Sometimes people died from fractures, either from infection or from not being able to protect themselves in an attack.

More than 2000 years ago, doctors in Egypt used strips of flexible rush called papyrus to straighten broken noses. Doctors in medieval England used splints of elm bark or wood to immobilize injuries. They also recommended hot baths and hot beer to prevent bones from shortening as they healed!

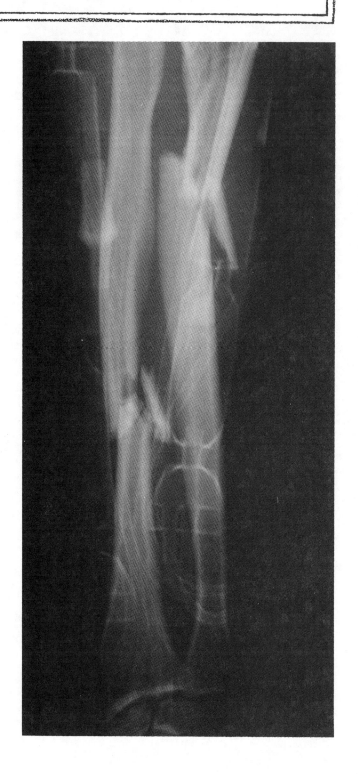

Here is an X-ray of the left arm of a 35-year-old man who has broken his ulna and radius bones. (In the background you can see intravenous tubes.) The bones are so badly broken that they have to be set with metal rods.

The first cast

During Napoleon's famous battles in the 1800s, surgeons on the battlefield discovered that bones healed better if the injured limbs were wrapped in bandages soaked in plaster of Paris. Casts like these are still used today. But doctors are just beginning to realize that sometimes they are too good at immobilizing bone. The problem is that any muscles and joints in the cast don't move either and they become quite weak. So with bad breaks or joint fractures, doctors use stainless steel pins and rods to hold broken bones together. A rod is placed on one or both sides of the limb. Pins are then inserted through the skin to the bone and attached to the rods. This device holds the bone completely still but allows surrounding muscles and joints to move. Another method for setting badly broken bones is to place a rod right in the bone to help it heal. Sometimes stainless steel plates and screws are attached to the bone (underneath the skin) to hold the fragments in place.

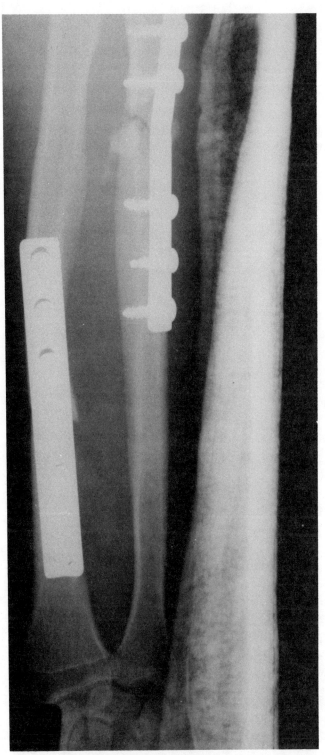

Here is a close-up of the same broken bones after the metal rods have been bolted in place.

Famous fractures

Sometimes when you cut yourself, the cut heals completely without leaving a scar. It's the same with bones. They can heal so well that even an X-ray won't show where the break was. But your skin can be scarred and so can your bones. Like your skin, bone is a living tissue and many of life's events, such as injury and illness, are recorded on it.

Dr. David Livingstone, the famous African explorer, was once mauled by a lion in Africa and his upper arm was broken. When he died, his remains were shipped back to England and that fracture line in his arm was one of the clues used to definitely identify his body.

BONE UP

If you've ever broken a bone, then you know how painful it is. Bones have nerves or pain "receptors" so they do hurt when you break them. But your skin and muscles are much more sensitive, so it's likely the damage to these "soft tissues" is what really gives you pain.

What hurts when you bang your "funny bone?" Your ulnar nerve. It's very close to the surface of your elbow and wedged tightly against bone, so when you bump it, it can't move out of the way. OUCH!

First aid for broken bones

What do you do if a friend breaks an arm or leg? The first thing, of course, is to get some help. But what if you are a long way from help? If that's the case, here are some important things to know when someone has broken a bone.

How do you tell if bones are broken?
- The arm or leg is lying at an unusual angle.
- The person is in pain and finds it very hard to move the limb.
- The injured area is swollen.
- The ends of the bone may have pierced through the skin.

What do you do if you can't get help right away?
- Don't ever move a person who is unconscious.
- Don't move anyone who is complaining of pain in the neck or anyone who feels he can't move.
- Try to keep the injured person warm and comfortable.
- Send someone for help.
- If you are alone, go for help and call an ambulance.
- Cover any exposed bones with a clean cloth.

THOSE AMAZING FLYING MACHINES

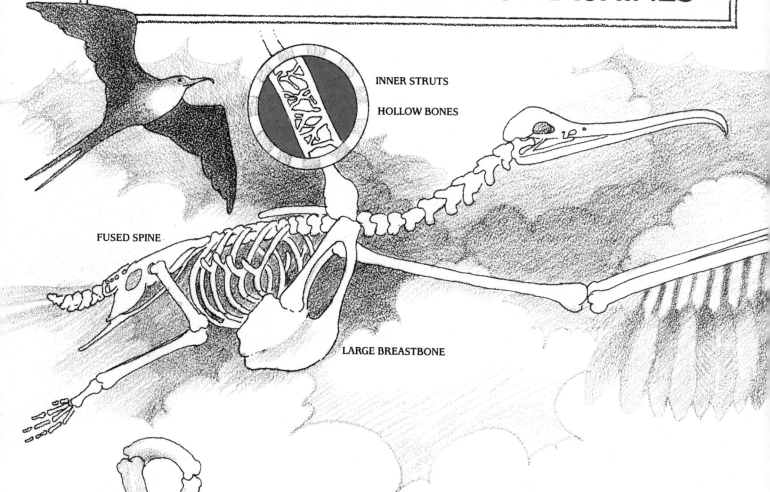

PEOPLE HAVE ALWAYS WANTED TO FLY. You've probably seen photos of those funny flying machines people built early in this century, or watched movies of brave inventors jumping off cliffs with huge, feathery wings or whirling propellers attached to them.

We now know that birds have more than just wings and feathers going for them. Probably the main reason why birds can fly is that their bones are extremely thin and light compared to ours. Some of their bones are actually air filled. There's even a passageway that connects the humerus (arm) bones directly to the lungs so that air can flow into the bones to give the bird an extra lift—a little bit like having an extra gas tank on board. Because bird bones are so thin and hollow, they need inner struts or supports to strengthen them. These supports are very much like the struts used in airplane wings.

Birds have many other ways of adding strength to their bones. For instance, since their wing feathers are supported by their arms and fingers, the finger bones are fused together to give them the strength they need. A bird's spine is fused for extra strength as well. A simpler way of adding strength is by

FUSED FINGER BONES

using a large bone. A bird's breastbone is very large, providing a place for powerful flight muscles to attach.

Look at the immense wing span of the frigate bird. It's hard to believe that its skeleton weighs only 112 g (4 ounces) or about as much as half a pear. Its entire skeleton weighs less than its feathers!

Does the frigate bird remind you of kites you've seen? There are good reasons why kites, airplanes and birds resemble each other. To get airborne, they all must combine some of these characteristics: lightweight, strong, streamlined and powerful.

There is a bird of prey called a "kite" and "kiting" refers to the way this bird swoops and soars through the air. Like the bird, the kites you make even have "spines" or "backbones." The kite's tail helps give the kite balance as it does in a bird. A bird can use its tail like a rudder, too.

You'll never have the hollow bones of a bird and so you'll never be able to fly, but as people have done for thousands of years, you can build a kite. The traditional diamond-shaped kite you can make on the next page looks a lot like a vulture swooping and soaring through the sky.

41

Make a flying machine

You'll need:

- a strong stick, .6 cm x 91 cm (¼ inch x 36 inches) (this forms the spine)
- a flexible stick, .6 cm x 86 cm (¼ inch x 34 inches) (this forms the bow wing)
- a small wood saw (have an adult help you use it)
- a pencil
- a roll of kite string [9 kg (20 pound) test] or fishing line
- all-purpose or fabric glue
- strong tape
- scissors
- lightweight cotton or nylon fabric, 97 cm x 102 cm (38 inches x 40 inches)
- a small 2 cm (¾ inch) diameter metal or plastic ring
- markers or paints

1. Make a 1 cm (½ inch) deep cut in both ends of each stick with the saw.

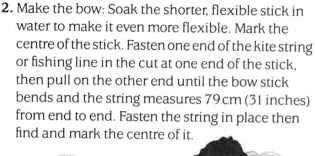

2. Make the bow: Soak the shorter, flexible stick in water to make it even more flexible. Mark the centre of the stick. Fasten one end of the kite string or fishing line in the cut at one end of the stick, then pull on the other end until the bow stick bends and the string measures 79 cm (31 inches) from end to end. Fasten the string in place then find and mark the centre of it.

3. Attach the spine to the bow stick: Place the spine stick on the centre of the bow stick so that the bow stick sticks out 18 cm (7 inches) above the bow string, as shown. With a small piece of string, tie and glue the bow string to the spine stick. Tape the bow and spine sticks together as shown. Run the line around the outer edges of the kite. Cut and tie the string at one of the ends.

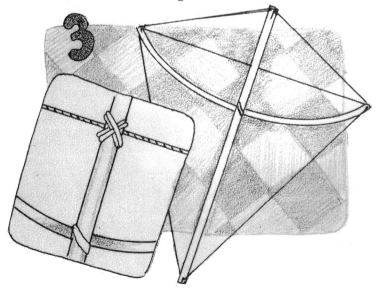

4. Cover your kite: Lay the skeletal frame on the fabric. Cut the fabric around the kite leaving 6 cm (2½ inches) extra material around the edges as shown. Fold the flaps over and glue them down. When the flaps are dry, untape the bow stick and push it up towards the top of the kite until it fits tightly against the fabric.

5. Attach the bridle and tow line: Tie the bridle and tow lines to the ring as shown. That's it. This kite should be very stable in flight—it doesn't need a tail. But you can add a small one if you like. Decorate your kite and fly it.

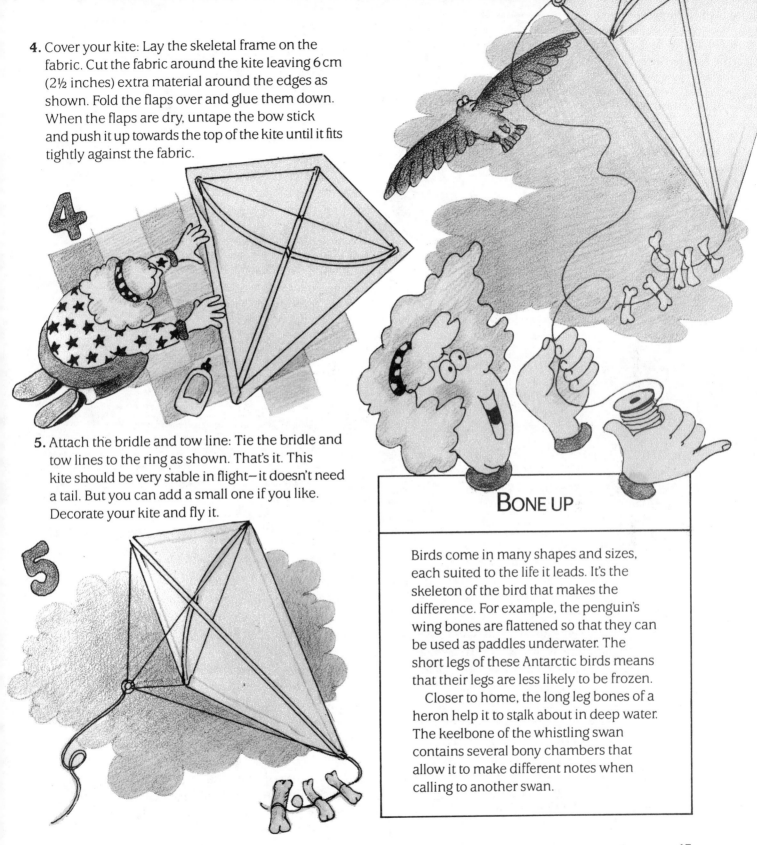

Bone up

Birds come in many shapes and sizes, each suited to the life it leads. It's the skeleton of the bird that makes the difference. For example, the penguin's wing bones are flattened so that they can be used as paddles underwater. The short legs of these Antarctic birds means that their legs are less likely to be frozen.

Closer to home, the long leg bones of a heron help it to stalk about in deep water. The keelbone of the whistling swan contains several bony chambers that allow it to make different notes when calling to another swan.

Who needs bones anyway?

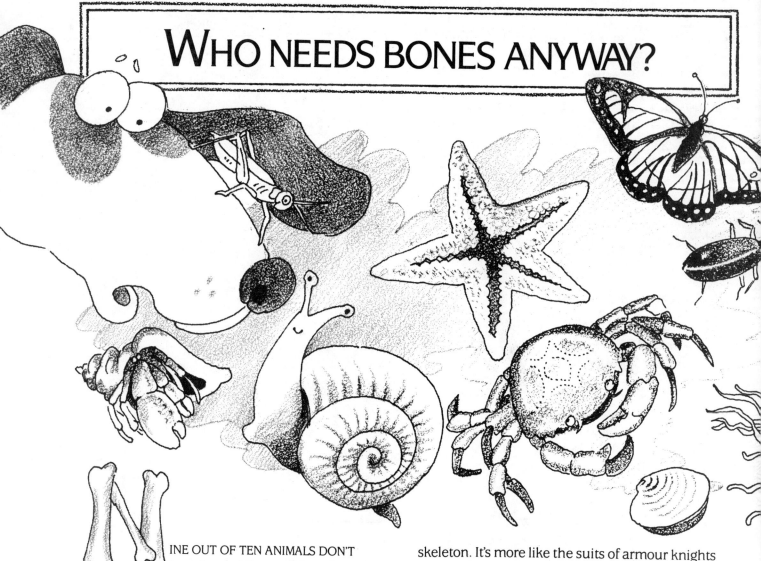

NINE OUT OF TEN ANIMALS DON'T have bones! Some of these boneless wonders are soft and squishy, such as worms and jellyfish, or hard and crunchy, such as beetles and clams. These animals don't have backbones so scientists call them invertebrates. (The word comes from vertebrae, the bones that make up backbones.) But in addition to not having backbones, these creatures have no bones at all! You've been reading about how much you need bones to protect your body and to help you move, so how do these boneless creatures survive?

Insects, for instance, wear their skeleton on the outside. This hard coat is called an exoskeleton (you have an endoskeleton) but it really isn't like your skeleton. It's more like the suits of armour knights wore long ago. The stuff this armour is made of is called chitin and it's a bit like your fingernails.

Another way you differ from insects is that your muscles are attached to the outside of your skeleton, while an insect's muscles are attached to the inside. This makes you much more flexible than a beetle, although having a suit of armour can have its advantages. Imagine the football games you could win and the sharks you could wrestle! On the other hand, an insect has to shed this armour every time it gets bigger. Each new suit is heavier and thicker than the last, which is one reason why animals with exoskeletons usually stay quite small or they'd be too heavy to move.

How do creatures with no bones move about? Bugs walk around on legs stiffened with chitin, but worms have to find another way. Most grip the ground with bristles, but that's not enough. Worms are a bit like long, thin balloons. If you squeeze one end of a balloon, it bulges out the other end. The earthworm moves by contracting muscles that make it longer and thinner, thereby pushing itself forward.

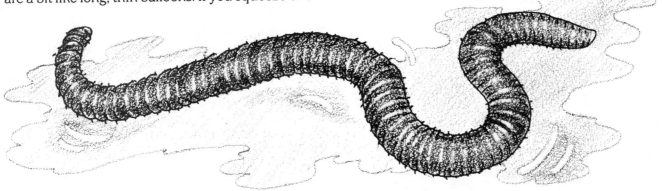

Scallops move by rapidly opening and closing their shells, then pushing water out behind them as they scoot along. A squid draws water into a special sac in its body, then shoots the water out again and away it goes. It steers by changing the direction the water jets out.

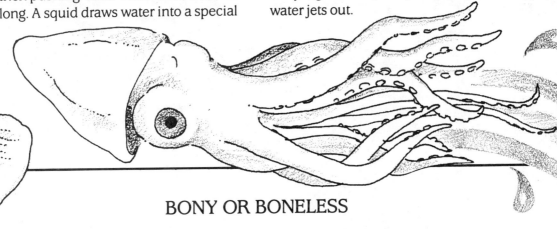

BONY OR BONELESS

Sometimes it's hard to tell if an animal has an exoskeleton or not. See if you can match each of these clues with the correct animal.

1. Like all insects, I'm hard on the outside and soft on the inside.
2. Some people might say I'm cranky. When I want to grow, I have to crawl out of my covering and hide under a rock for a few hours or days until my new covering is hard.
3. I can tuck my head right inside my shell. My shell grows with me.
4. I pick up my hard cover on the beach. Glass, shell, rock — I stick them all on my back.

a crab

b turtle

c hermit crab

d beetle

Answers on page 96.

A BONE-HARD PUZZLE

SOME THINGS LOOK LIKE BONE, BUT AREN'T. Other things don't look like bone but they are. See if you can decide what's bone and what isn't in this puzzle by using what you know about bone. You already know:

- bone is alive and growing
- it bleeds but it can heal itself
- it's very strong and it's made from collagen and calcium phosphate.

1. These magnificent antlers grow out of the frontal bone of the skull. Antlers drop off during the winter and a new, larger pair grow in the spring. These new antlers have blood vessels and nerves. They even get itchy. Are they bone?

2. Horns are found on many animals—sheep, goats, antelope and cattle. Many years ago, a zoo received a shipment of horned animals from South America. To save space, the shipper had cut off the animals' horns, thinking they would grow back. They didn't and the zoo was left with some odd-looking animals! Perhaps if the shipper had known that horns don't have nerves or blood, he wouldn't have cut them. Are horns made of bone?

3. These magnificent ivory tusks sure look like bone. They are incisor teeth that never stop growing. Teeth are made of calcium minerals and dentine but no collagen. Are teeth bone?

4. An armadillo is certainly a well-armoured mammal. Its back, sides, head, tail and legs are covered with small, strong plates that grow as the animal grows and keep it protected. Some armadillos can roll into a ball to shield their soft bellies. Do they have bony plates, china plates or something else?

5. The turtle's shell covers most of its body. Part of it actually grows out of the vertebrae. Could a turtle's shell be bone?

6. Growing at the end of the stingray's tail is a single spine. It's long and serrated like a bread knife and a gland at the base contains a powerful poison. Is this spine bone?

7. Some corals reefs look like antlers. They are very hard but are made up of soft-bodied creatures. Is coral bone?

8. The Inuit used baleen to make fishing nets, baskets and snares because it is so elastic. In Europe, it was used to make corset stays and whips. Baleen looks quite soft and hairy, but it's sometimes called whale bone. Hmmm… very confusing. What's your guess?

Answers on page 96.

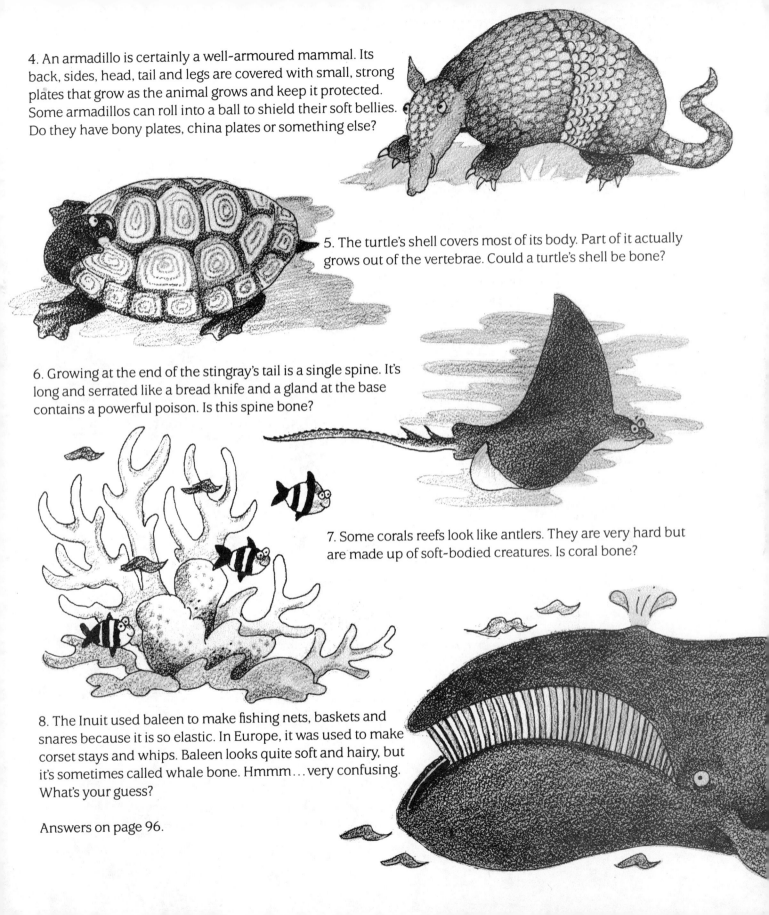

Tales bone tell

How do we know that some dinosaurs were a little clumsy or that dogs have been our pets for thousands of years? Were Neanderthals actually kind, intelligent people? Did Egyptians have pet gerbils, thousands of years ago? These stories and more are mysteriously hidden... in bones.

DIGGING UP OUR PAST

MUCH OF OUR story—the story of people on planet Earth—can be read in our bones. The story began more than 12 million years ago when a group of apes began changing in ways that would one day lead to *Homo sapiens* or humans. You can see for yourself how bones can tell us about this long process of evolution.

There are many theories to explain evolution. One is that a cooler, drier climate reduced the number of forests in which apes could live. When apes began spending more time in grasslands and less time in trees, their bodies slowly began to change. Why? Because to survive on the grasslands, you had to see danger coming your way. So it was time to stand up to get a good look. Standing up wouldn't have been easy at first. But over thousands of years, the pelvic bones changed to allow our ancestors to walk upright. An upright posture also freed their hands for carrying food, tools and babies.

What was life like for early people? Did they eat mammoth burgers for lunch? Archaeologists are always sifting the ground for more answers to these questions. Let's look back 40 000 to 100 000 years ago to the time of our ancestors known as the Neanderthals.

Lots of people have the idea that Neanderthals were stupid people who went around hitting each other with clubs. We know from their thick, heavy bones that they were short and muscular, and their skulls tell us that they had big noses, receding chins and large brow ridges over their eyes. Maybe they didn't look too bright, but there was enough room in their skull for a brain as big as ours.

We know that some Neanderthals lived in huts made of animal skins stretched over a frame of mammoth bones. We also know that Neanderthals dined on mammoth, woolly rhinoceros, deer, horses, elk and wild goats because bones of all these animals were found buried with the blackened remains of campfires.

There is evidence that Neanderthals took care of each other. Skeletons of hurt and disabled people, such as one with a deformed leg and foot, tell us that they must have been cared for in the community or they wouldn't have lived to adulthood.

Neanderthals were also the first people known to bury their dead. Sometimes they put flowers in a grave and placed animal bones around the grave's edge.

What became of Neanderthals? They may have stopped evolving and died out completely or they may have evolved into modern humans. Anthropologists don't know for sure and are always on the look-out for bony remains that will help them solve the mystery of our evolution.

Look carefully at the skeletons shown here. One is an ape, one represents a step forward in evolution (*Homo australopithecus*) and one is *Homo sapiens* – us. What differences can you spot between them? For example, the shape of the brain case is different for each. Look also at their arms and legs. There are at least five differences between the skeletons. How many can you find?

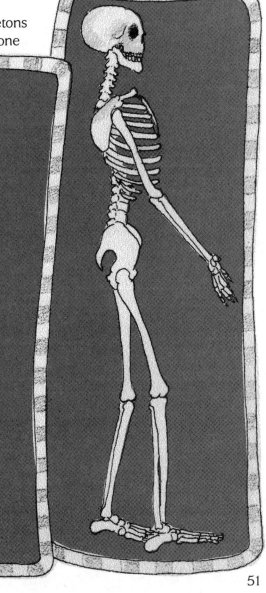

Answers on page 96.

Time Travellers

IF YOU'VE EVER WANTED TO TRAVEL BACK IN TIME TO see how people lived 1000 years ago, here's your chance. Climb into the Discover Time Machine and step back to a time before the first Europeans came to North America, to the time of the early Iroquois culture.

It's early morning. There's a heavy mist blanketing the ground, so you walk slowly, expectantly.

The villagers are just beginning to stir. A dog yelps, perhaps sensing your presence...

You glance at the ground and suddenly realize that you're no longer wearing shoes. Your clothes have changed too: they're now made of animal skins, and a necklace made of bone beads hangs from your neck.

As the sun burns through the mist, you begin to see the village more clearly. A short distance away, a woman is decorating a pottery vessel. She's poking a small blunt bone into the clay along the rim of the pot. Little bumps, like small peas, appear along the edge. Then she picks up a piece of notched bone and pushes it several times into the clay.

Nearby, a young woman is making clothing. She's punching holes in leather with a very pointed bone. Some bone needles and strips of leather lie on the ground beside her. You realize this must be how your clothes were made.

By the river, a man is carving with a most intriguing looking tool. It looks like an orange tooth attached to a bone. You notice that all the tools you see are made from animal bones—no iron or steel here. The notched pottery decorator the woman is using is actually a piece of turtle shell; the leather puncher or awl is made from a catfish backbone; the carving tool is a beaver tooth attached to an antler. Each animal bone has just the right strength and shape for its job.

Practise prehistoric pottery

Look closely at these clay pots. How many different designs do you see on them? These designs were all made with the bone and wood tools of the Iroquois. Hollow bird bones, carved bones, sticks or bones wrapped with cord, and turtle shell were used to create these designs. See if you can recreate some of these patterns yourself in your own clay work.

You'll need:

○ clay
○ a rolling pin
○ a kitchen knife
○ a fork
○ a stick or bone and string
○ a drinking straw or small hollow bone

First, make a small pot out of clay. Here are two methods you can try.

1. Put your thumb in the centre of the clay and turn the clay slowly around your thumb, pinching the clay upwards as you turn it so that it forms the sides of the pot.
   OR
2. Flatten the clay using a rolling pin. To make the sides of your pot, cut a long strip of clay that is as wide as you want your pot to be tall. Sprinkle water on the two short ends of the strip and join them together in a circle by gently squeezing and smoothing them with your thumb. Cut a round bottom to fit and join it to the edges in the same way.
   A. Make small pea-like designs along the rim of the pot by pressing a stick or bone into the inside rim. With your other hand, support the outside of the pot to prevent the stick from piercing right through. Archaeologists call these designs punctate.
   B. Make designs on your pot by experimenting with different objects. Try dragging a fork across the clay. Wrap string around a stick or bone and press or roll it into clay. Press the straw or hollow bone (with the end sawed off) into the clay.

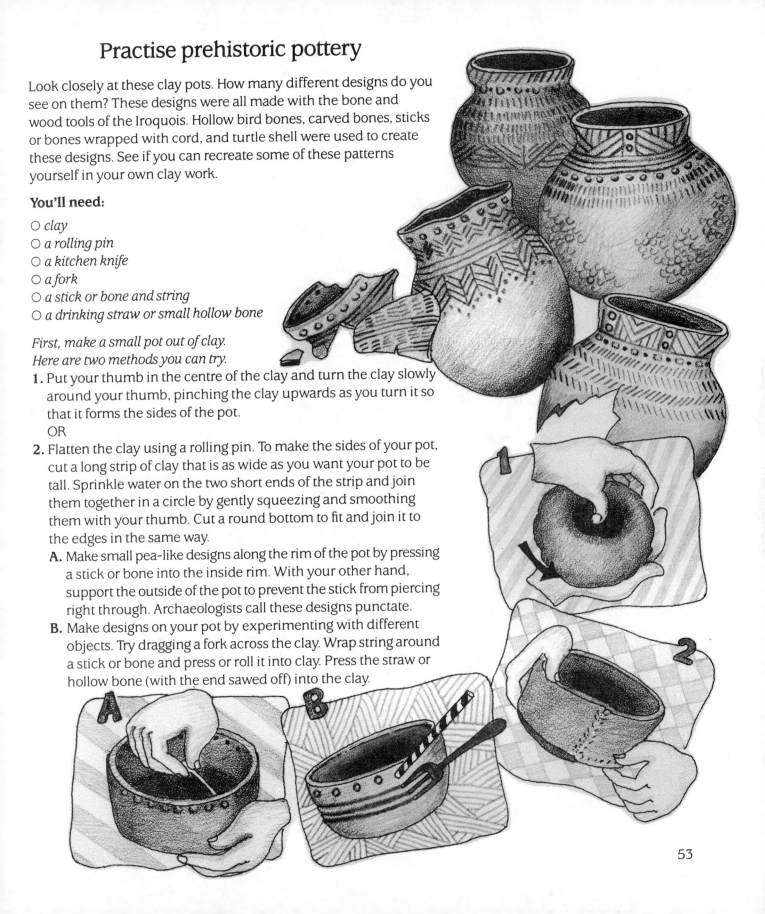

Dogs and their Bones

ARCHAEOLOGISTS IN ISRAEL HAVE discovered the 10 000-year-old skeleton of a man buried with his hand resting on a puppy. These bones are evidence that there has been a close relationship between dogs and people for a long time. In fact, sometimes it looks like the bond between an owner and dog is so strong that they sometimes look a bit like each other.

Actually, we all resemble dogs! What? Look beneath the fleas, the fur, the slobbery tongue and the limited vocabulary and what you see is a skeleton remarkably like your own. In fact, all mammals (warm-blooded, milk-giving creatures) have the same type of bones (give or take one or two).

Still don't believe you really have the same bones as your dog? Compare the human skeleton here to any of the dog skeletons. How many similarities can you find?

BASSET HOUND

Look closely at the ribs, shoulder blade, hip bones and "arms" of both skeletons. Sometimes it's hard to tell your bones from the dog's.

Now look at the dog's tail. It's actually a continuation of the dog's spine so it has vertebrae bones in it, too. A long tail contains more vertebrae than a short one. If you reach back and feel at the bottom of your spine, you'll notice a knob of bone, the remainder of a tail we used to have.

One big difference between dogs and humans is our skull. A dog's jaw is usually thick and heavy—just what's needed for gnawing on bones and raw meat. Now take a look at those teeth. Since dogs are carnivores, they need large, pointed teeth for slicing through their meaty diet. Our teeth are suited to our mixed diet of meat, fruit and vegetables. But both dogs and humans have a mixture of different teeth and that sets them apart from reptiles, whose teeth are all the same.

If you look at the skeletons of polar bears or grizzly bears or any other species of bear, you'll see that they all look fairly similar. But check out the skeletons of different breeds of dogs and it's a different story. Over the centuries, dogs have been bred to do different jobs. Some dogs were bred to be long and thin (so they could go down burrows after prey), others strong and gentle, others fierce and aggressive.

Now look at the skeletons of the dogs shown here. Which one do you think was bred to be:

- a slender, speedy racer?
- a hound with short legs and a keen nose for sniffing out prey?

Answers on page 96.

BONE UP

Why do dogs bury their bones? Probably for the same reason squirrels bury their nuts. They are "saving" their food for a time when there might not be much available. They are also hiding their bones from other dogs and animals. The only problem is that dogs often can't find their bones again!

GREYHOUND

Get digging!

EVEN IF YOU CAN'T GO TO AN archaeological dig, you might be able to a little digging in your own yard. Maybe your parents have construction projects planned that require some excavating. Perhaps they're thinking of making a sandbox pit, turning the earth to make a new garden bed or building a new deck or basement entrance.

These are great opportunities to get below the ground and see what you can find. If there aren't any construction projects planned, you could ask if there is a small area in your yard where you can dig. A good place would be right at the back where there might have been an old garbage dump.

Bone up

Garbage heaps are great sources of information. When archaeologists excavate these, they always examine the contents very carefully. What do you think a scientist would learn sifting through your garbage? The bones in your garbage would reveal whether you had chicken or beef for dinner recently, how old the animals were when they died, how they were cut up and possibly how they were cooked.

Dig down

You don't need to dig with the precision of an archaeologist but you can use some of their techniques.

You'll need:

○ *permission from your parents*
○ *gloves*
○ *a shovel*
○ *a trowel*
○ *a screen*
○ *buckets of water*
○ *an old toothbrush*
○ *a notebook and pencil*

1. Use the shovel to dig a hole about 20 cm (8 inches) down and 50 cm (20 inches) across.

2. Look for any objects in the soil you removed and any objects sticking out of the sides of the hole. Use the trowel to carefully remove them.
3. Gently sift the soil you removed through a screen.

4. If you see something interesting, wash it off in a bucket of water. Use the toothbrush to gently clean it.

5. Try placing some of the soil in a bucket of water. This will cause very small bones to float to the top.

6. Make a sketch of anything interesting you have found. Note where you found it: for instance, 20 cm (8 inches) down, near the left-hand side of the hole.
7. If you've found something you think your local museum would like to see, call and speak to an archaeologist or other expert who works there.
8. Dig down another 20 cm (8 inches) and see what you can find.

Other places to look for bones

Sometimes archaeologists and palaeontologists just get lucky and find artifacts (old objects made by humans) or bones lying right on the ground. "Ah," they think, "maybe there are more underneath." Ploughed fields often turn up bones and artifacts. Sometimes if topsoil looks extra rich in one spot, it indicates that there is a garbage dump from an ancient culture underneath. Garbage dumps, or middens, along river banks are sometimes exposed by erosion. Places where trees have blown over and roots are exposed can be good places to look, too. But why would an archaeologist examine the area around a groundhog hole? Even animals dig up artifacts when they make their underground homes!

What you might find:

In pioneer times, lots of things were made from bone. You might find bone toothbrushes, bone buttons, bone button hooks (a tool for doing up buttons), bone bottle tops, bone necklaces or bone dice.

A lot of the bones you find may have been buried by dogs. You can tell because:
- they look like the kind of bones found in roasts, steaks or chops.
- they may have a flat surface with marks made by a butcher's saw.
- they may have teeth marks on them.

At one time much of North America was under a great sea. You might find fossils of sea creatures or sea shells in the rock.

Have you ever buried a pet in your back yard, maybe a hamster, gerbil, bird or cat? People that lived in your house before you did may have buried their pets, too.

You might find part of a bird, squirrel or mouse skeleton. How will you know?
- bird bones are generally quite tiny and hollow.
- mouse or beaver skulls usually have four large teeth (incisors) in front. (See page 60 for an identification key that can help you identify your bones.)

BONES IN YOUR BACK YARD

Seen any archaeologists digging around in your back yard lately? It has happened! Find out more from Mima, a museum archaeologist.

"You might be wondering why I want to dig up back yards! Well, let me describe the two back yards I've been excavating. They both back onto a ravine that has a freshwater spring draining into a creek. Since springs often have provided water for animals and humans for thousands of years, they make good archaeological sites. With permission from the homeowners, I start digging. For my next project, I'm planning to dig up front lawns. Really I am! If you get a chance to dig in your back yard, you may unearth some great artifacts. You can let your local museum know if you think you've found something important."

DIG LIKE AN EXPERT

When archaeologists start digging, they do it very slowly and carefully so that they don't destroy important evidence. Imagine attacking a huge hill of dirt with a toothbrush and dental pick, and you'll get some idea of the patience you need to be an archaeologist. Of course, archaeologists use shovels and even bulldozers if they are sure they won't damage anything.

Where bones and artifacts are found is as important as what is found, so archaeologists make sure they meticulously record the location of all the items they find. They divide the surface of the site into squares with stakes and string. The objects they find are then mapped on a piece of paper that has been divided up using the same grid system.

Underneath you are layers and layers of soil, each laid down over the centuries. So something found 1 m (3 feet) down is probably much older than something found just 5 cm (2 inches) under the ground.

Examine an owl pellet

If you live in the country or even some parts of a city, you might even find an owl pellet. It looks like a compact wad of hair and bones. You'll find owl pellets under trees or in barns where the owls roost. These pellets are sometimes called "mouse kits" because they may contain the skeletons of several mice.

You'll need:

○ *a small knife or darning needle*
○ *a magnifying glass*

1. Soak the pellet in water, then gently break it open.
2. Separate the bones from the fur with your knife or needle.
3. Examine the bones with the magnifying glass and put similar bones together in a pile.
4. Compare the bones to the ones shown here. Can you match them?
5. What do the bones tell you about the owl and what it eats?

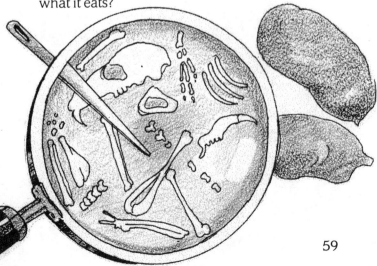

Bone Detectives

IF YOU GAVE A PALAEONTOLOGIST A BONE FROM a mouse or a bird, chances are she'd be able to quickly tell you what animal the bone came from. Amazing! You too can have this secret knowledge! The next time you find an animal skull, try identifying it with this simple identification key.

First, check your skull for teeth. If it has none, look below at No. 1. Look at No. 2 if it does have teeth. Keep reading the descriptions until you find the one that fits the skull you have.

1. ANIMAL HAS NO TEETH:

a. Skull is heavily built, with short broad jaws and no obvious beak: turtle

b. Skull is lightly built, often with many thin parts and an obvious beak: bird

2. ANIMAL HAS TEETH:

a. Teeth are small and all of the same general shape: reptile or amphibian

- the skull is broad and flat with large eye sockets and many tiny teeth: frog
- the skull is made up of many long, thin, rather loose bones; the teeth are sharply pointed: snake

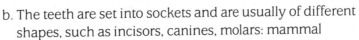

b. The teeth are set into sockets and are usually of different shapes, such as incisors, canines, molars: mammal

- there are one or more molars in each jaw, enlarged for tearing meat; the canine teeth are usually long and pointed: meat eater such as a wolf
- several cheek teeth in each jaw are nearly the same size and shape; the canine teeth are usually small, or absent: herbivore, such as a deer
- front teeth are usually enlarged and chisel like; there are no canines: rodents such as a beaver or muskrat

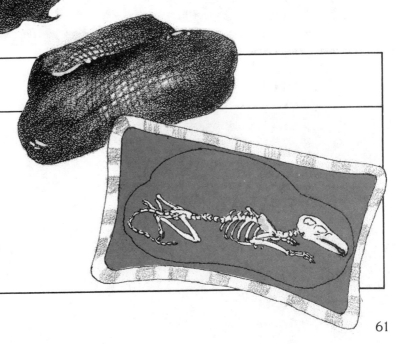

BONE UP

What do you think this black lump is? It turned out to be a mummified gerbil equipped with a little sack of food for its after-life. By being able to identify animal bones, archaeologists can tell what animals were mummified in ancient Egypt. These include crocodiles, birds, cats and this little gerbil.

Finding Fossils

WHEN YOU'RE digging in your back yard, you may be lucky enough to uncover a fossil. Fossils are what's left of any ancient animal or plant. They can be very old bones, teeth, shells or even the impressions of things in rock or earth, such as ancient sea creatures or dinosaur footprints. Usually only those plants and animals that are buried quickly before they begin to decay become fossils. When a bone is buried for millions of years, its pores fill up with minerals, usually silica or calcium. Because of all these minerals, a fossilized bone may look and feel like stone but some of the original bone will still be there.

How do fossil hunters know where to look? Places where the land has been eroded away by wind, rain or water, such as the Alberta badlands or sea cliffs, are good places to look for fossils.

You might not be able to go on a fossil hunt right away, but museums are great places to find out more about fossils. Your local museum might have displays that explain what fossil collectors do, and they may even offer fossil-hunting field trips.

You can find invertebrate fossils (such as ancient sea creatures) throughout North America but if you're looking for vertebrate fossils, you're most likely to find them in the badlands of western North America, in the large areas of exposed sedimentary rock.

Bone up on the following tips from the experts on looking for vertebrate fossils.

- Fossil hunt with an adult.
- Wear sturdy boots and protective clothing.
- Where possible, find out what extra precautions you should take, such as carrying extra water, looking out for snakes and watching for flash floods.
- Be extra careful around cliffs, overhangs, loose sedimentary rock, river banks and lake shores.
- Always get permission to look or collect if you are in a park or on private property.
- If you think you've found a partially buried fossilized bone, let the park authorities or local museum know. They may want to excavate it and often they will let you help.

To learn as much as possible about the bone, it's important that it be excavated carefully.

- In some places, you are allowed to collect any loose bones lying on top of the ground. Check first.
- A fossilized bone will likely be harder and heavier than you'd expect. All those extra minerals can make a fossilized bone ten times heavier than the original bone!
- A bone that has been exposed to sun, rain and snow for a few years will look white, worn, cracked. It might look very old but it may not be. Don't let it fool you. Even if it's not old, you can still try to identify it.

Excavate a cookie

Getting a fossil out of a rock is a little like getting a chocolate chip out of a cookie. Try it!

You'll need:

○ one or more chocolate chip cookies
○ a small pointed instrument such as a skewer or tiny screwdriver
○ a table knife

1. Don't eat the cookie.

2. Choose a chocolate chip to remove. "Excavate" around the edges of the chip by gently tapping the end of the skewer with the handle of a knife. Keep tapping until the chip comes free. Could you do it without damaging the chip? It might take practice.

3. Now imagine that each chocolate chip is a bone that has to be excavated very carefully, and its exact position in the cookie recorded. Is it any wonder that excavations of bones can take years!

4. Eat the cookie!

BONE UP

When scientists discover fossilized bones, they are very careful to remove them from the ground without damaging them. The bones and the surrounding rock are then encased in plaster before being moved back to the museum. Once there, they are cleaned and coated with a thin film of plastic solution that holds any loose fragments of bone in place. If the bones are part of a skeleton to be displayed, replicas will be made, then drilled and fitted with steel rods. These rods are then welded together to hold all the bones in position.

READ ANY GOOD BONES LATELY?

IT'S AMAZING HOW MUCH WE KNOW about animals that died more than 65 million years ago...and all from their bones. If you wanted people a million years from now to know what your life was like, what would you tell them? Would you write about your family, your dog, your terrible cold, the home run you hit, or what you liked best to eat? Even though dinosaurs and other ancient creatures didn't leave us any books about their life, they left us many clues. Bony clues. We read their fossilized bones and teeth instead of their diaries.

You probably already know all sorts of amazing things about dinosaurs. But did you know that...

• the smallest dinosaur palaeontologists have found was about the size of a chicken! By examining the bones of *Compsognathus*, experts estimate that this dinosaur was only 60 cm (2 feet) long and weighed less than 1 kg (2 lb.). Its bones were so delicate that they have been crushed while being fossilized but scientists have still been able to reconstruct this tiniest of dinosaurs.

• there were dinosaurs more than twice as big as *Tyrannosaurus rex*? *Ultrasaurus* was 16 m (52 feet) tall, 30 m long (98 feet) and may have weighed 130 tonnes (128 tons). But wait! The tail of a brontosaur has been discovered that is sooooo big it's being called *Seismosaurus*— "Earthquake dinosaur"! By measuring the tail bones, scientists can guess that the rest of *Seismosaurus* must be enormous, too. However, after two years of excavating, scientists have only uncovered as far as its hip bone, so no one knows for sure!

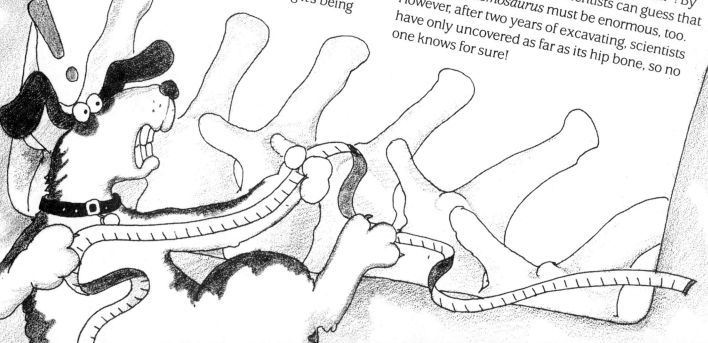

- Hadrosaurs may have been rather clumsy? It seems they often stepped on each other's tails! How could we possibly know this? One palaeontologist at the Tyrrell Museum in Drumheller, Alberta, noticed that many of their hadrosaur skeletons had cracks in the vertebrae of their tails. Because hadrosaurs travelled in herds, it looks like they might have been tailgating!

- palaeontologists think that *Coelophysis* may have been a cannibal. The bones of two young *Coelophysis* were found in the rib cages of two adults. Some experts suggested that they might be the bones of unborn babies but most scientists agree that these smaller skeletons are too big and well-formed. Because skeletons of this predatory dinosaur are found in groups, palaeontologists believe they lived in herds, unlike any predator alive today.

- *Euoplocephalus* was such a well-armoured dinosaur that in addition to bony spikes and armour on its neck, tail and back, it even had bony eyelids! Its name means—what else?—"well-armoured head." Experts also think *Euoplocephalus* had a great sense of smell. Why? This dinosaur had looped bony air passages in its nostrils, rather than straight ones. That meant there was room for many sensory nerves that do the smelling. So *Euoplocephalus* could get a really good whiff of any approaching predators.

MUSEUM BLOOPERS

Scientists in charge of assembling dinosaur bones at museums have to know a lot about how joints fit together. But even the experts make mistakes. For instance, German palaeontologists in the early 1900s put a *Diplodocus* together the wrong way. They connected its thigh bone to its hip socket in a way that made its legs so short, its stomach would have dragged on the ground! If this *Diplodocus* had climbed down from its mount, it would have had to find deep ruts to walk in!

Build a dino

Assembling dinosaur bones can be tricky. So how do palaeontologists do it?

They often look at today's animals for clues. For example, when assembling a mammoth, they look to the elephant for help. In fact, if there weren't any elephants around (or frozen mammoths), it would

READ SOME DINOSAUR BONES

Dinosaur bones can tell fascinating stories. The captions on the *Diplodocus* below describe what to look for on this skeleton or on dinosaur skeletons you see in museums. Read the information, then see what you can deduce about *Diplodocus*. Was it a meat eater? Can you see any evidence of broken bones on this skeleton?
Answers on page 96.

Size of limbs: Are the hind limb bones much larger than the front ones? This tells you that the animal was probably a meat eater since it was able to run fast enough to catch its prey.

Muscle scars: The shaft of a bone is smooth unless there is a scar where the muscle was attached. The scars look like rough ridges near either end of the bone. A large ridge may indicate that a powerful muscle was attached there.

have been very hard to figure out that mammoths had long trunks.

Once the bones are assembled, the palaeontologist looks carefully at the "muscle scars" on the bones. These raised, roughened areas show where the muscles were once attached. A big ridge or bump indicates where a large muscle was attached. Once the experts have examined the bones, models of the muscles can be made. But how do palaeontologists know what colour the dinosaur's skin was or how thick its legs were? Did its tail drag on the ground? Did the dinosaur have claws? Was its skin pebbly or scaly? These are challenging questions and often can't be answered for sure.

Broken bones: Look for lumps and raised, thickened areas on the shaft of bones where a break has healed. Breaks in the rib bones are the most common. It's unlikely that you'd find evidence of broken leg bones—the animal wouldn't have survived.

Size and shape of teeth: Are there long, pointed teeth (canines)? The animal was probably a meat eater.

Are the teeth mainly broad and flat? The animal was likely a plant eater.

Are the teeth a mixture of broad, flat and pointed? The animal probably ate a diet of plants and meat.

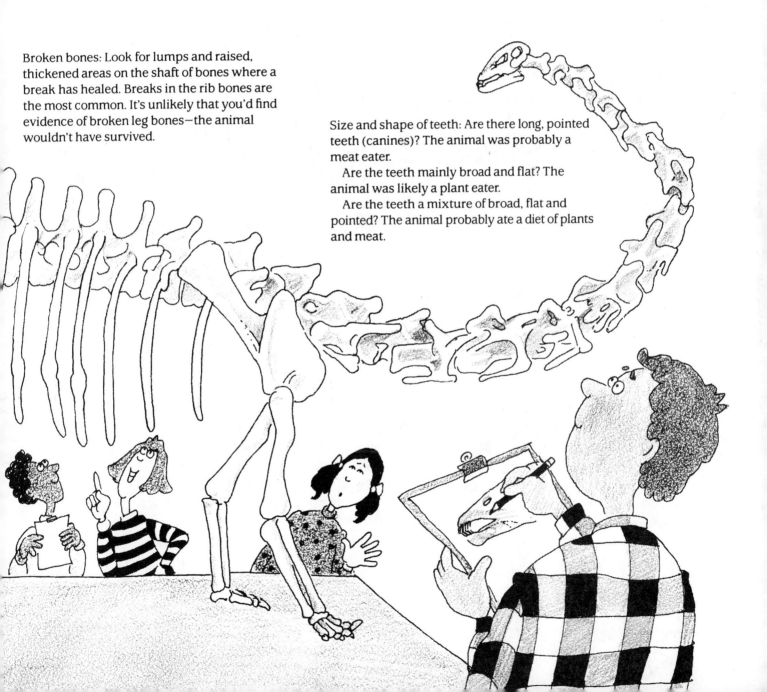

ONLY YOUR ARCHAEOLOGIST KNOWS FOR SURE

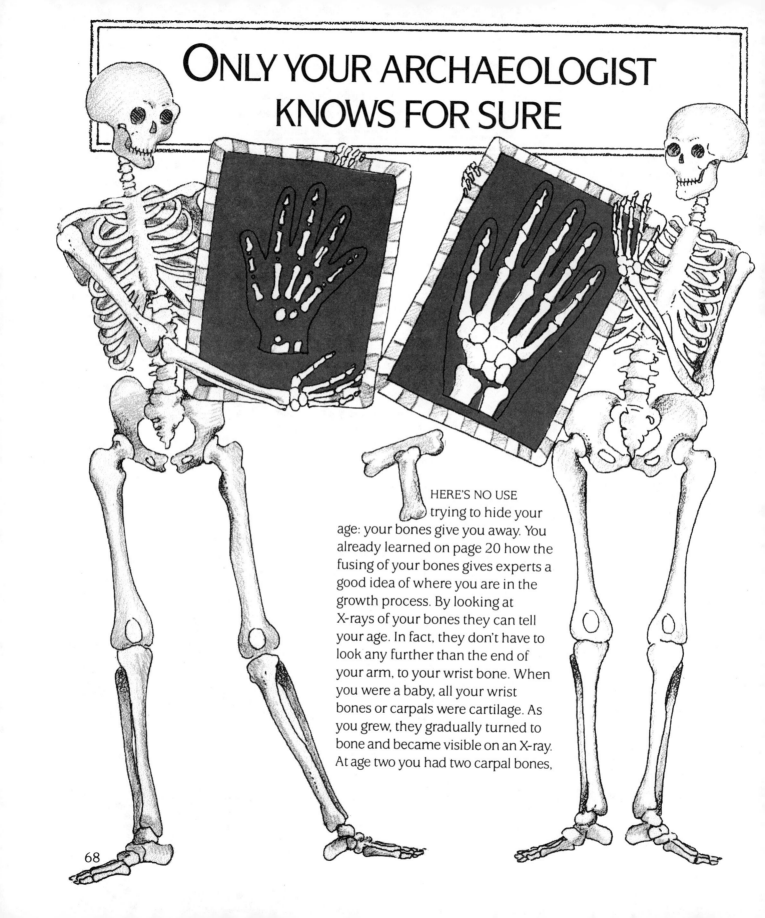

THERE'S NO USE trying to hide your age: your bones give you away. You already learned on page 20 how the fusing of your bones gives experts a good idea of where you are in the growth process. By looking at X-rays of your bones they can tell your age. In fact, they don't have to look any further than the end of your arm, to your wrist bone. When you were a baby, all your wrist bones or carpals were cartilage. As you grew, they gradually turned to bone and became visible on an X-ray. At age two you had two carpal bones, at age three you had three, when you were five you had five, at age eight you had seven and when you are 12 you have eight. In the X-rays the skeletons are holding on the opposite page, can you tell which belongs to a two-year-old? Answers on page 96.

Scientists can combine their knowledge of how bones form and fuse with some high-tech equipment to find out a lot about mummys. They don't even have to unwrap them! For instance, archaeologists can X-ray a mummy they're studying and if the ends of the bones haven't hardened together, they know it was a child. If some bones have fused but not others, then the mummy was probably a teenager. The size of the skeleton, and the number and condition of the bones and teeth, give them more clues about the exact age the person was when he died.

Boy or Girl?

Bones can also tell archaeologists if a skeleton belonged to a male or a female. You just have to know what to look for. Experts look at the thickness of the bones, the width of the pelvic bone and how the width of the pelvic bone compares to that of the shoulder bones.

Look at the two skeletons on the left then see if you can figure out which is which. It will help if you think about how the proportions of men's and women's bodies differ from each other.
Answers on page 96.

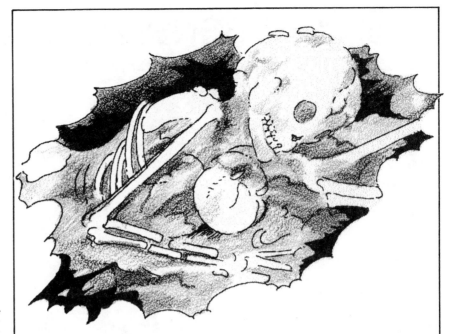

DISASTER AT HERCULANEUM

Two thousand years ago, Mt. Vesuvius in Italy erupted, spewing deadly volcanic ash and molten rock over the countryside. The 5000 people of the nearby city of Herculaneum were buried under 20 m (66 feet) of lava. By studying their skeletons, archaeologists are learning about their lives and their desperate last moments. For instance, one girl's bones revealed that she was about 14 years old and was used to carrying very heavy loads. The baby she was clutching wore gold jewellery, a sign that it was from a wealthy family and the girl was probably a slave who worked for the family.

The archaeologists also ground up small fragments of the bones they unearthed and analysed them. One thing they discovered was that some people had very high lead levels in their bodies from drinking and eating out of lead-lined cups and plates. The scientists also studied the way these ancient people's bones and teeth formed. They could tell that one woman had gum disease because the bone along her gum line was pitted with tiny holes. If she was alive today, she would be told to floss more often!

SPECIAL BONES FOR SPECIAL JOBS

SOMETIMES BONES tell the story of how an animal lives or once lived. For instance, knock on the top of your skull with your knuckles. There's less than 7 mm (1/3 inch) of bone you've got there protecting your brain. Now imagine another 23 cm (9 inches) of bone on top of that. That's just what the head banger dinosaur, *Pachycephalosaur* (meaning thick-headed or bone-headed), had on top of its head — a thick layer of extra bone to ram other head bangers with. Ouch! Woodpeckers have extra bony heads, too. All that hammering on trees could give them quite a headache if they didn't have extra thick skull bones to protect their brains!

Every animal has developed special bones that help it survive. For protection, this armadillo is completely covered with bony plates, right down to the tip of its tail.

And check out these two fish skulls. Can you guess which fish eats fruits and nuts and which one pokes its nose into tiny crevices to find food?

Answers on page 96.

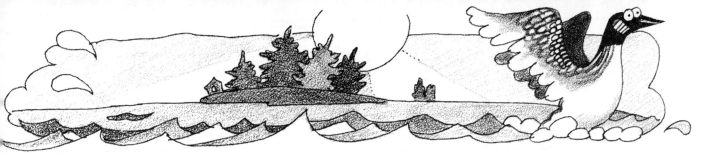

Have you ever noticed how a loon races over a lake's surface before it takes off? It's not sight-seeing. Most birds have hollow bones so that they are light enough to fly. But the bones of a loon are very dense, which helps this water bird dive deep in search of food. In fact, a loon is so heavy that it needs at least a 300 m (985 foot) runway to take off.

No one is ever going to call this snake spineless. With more than 200 vertebrae to its name, it's pretty well all backbone. The snake has developed a very special skeleton for life on its belly. Its flexible backbone allows it to slither along the ground in a series of curves. Actually, fossils show us that snakes' ancestors once had legs. But over millions of years as living conditions changed, legs were no longer useful, so they gradually disappeared.

The snake has found some interesting solutions to the problems it faces every day. For instance, what would you do if your dinner was larger than you and you couldn't chop it into bite-sized pieces? You could do as a snake does and swallow it whole. But how could you swallow something bigger than your mouth? Well, if you were a snake you'd do it with jaw bones held together by ligaments so stretchy that they allow your jaw to unhinge and your mouth to open very wide. Snakes that eat eggs whole also have sharp bony spikes in their throat that break the egg open when the snake bends its head down.

Find out how your own jaw bone works. Place your hands on your face just in front of your ears and slowly open and close your mouth. What do you feel? Move your bottom jaw side to side. Do you feel the hinges in your jaw move? Look in the mirror to see how the shape of your face changes. Then put your fingers on your upper lip and open and close your mouth. Can you tell if it's your top or bottom jaw that does most of the work? Answer on page 96.

YOU LOOK JUST LIKE... YOUR BONES

YOUR BONES TELL A LOT ABOUT YOU. To find out what they have to say, start by giving your scalp a massage. Can you feel bumps and grooves? Now move down to your face. Gently feel the corner of your eyes, then trace under your eyes and over to the bridge of your nose. Feel the ridge of bone under your eyebrows. You've just outlined your eye sockets. They can vary from large, which means you have big eyes, to quite small. Next feel your cheekbones. Some are high on the face, some are low, and some are very pronounced. What kind do you have? Now trace around your jaw. Some jaws are wide and square, others are narrow and rounded.

Now you know what police artists look for when they help detectives identify skulls. The detectives need to know how the person looked when he was alive and the artists use their knowledge to create an image. One way of doing that is to apply Plasticine directly to the skull to build up the facial features. But artist Bette Clark uses a computer to discover what the person looked like. Here's what she had to say about the process.

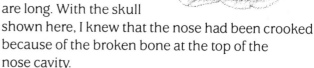

"At first, all the skulls looked the same to me— a skull was a skull. But I began to see that the bones were unique. For instance, some nose cavities are wide, some are long. With the skull shown here, I knew that the nose had been crooked because of the broken bone at the top of the nose cavity.

"The eye sockets were quite large, so I drew in large eyes. Then I noticed that the teeth stuck out

and I knew that that usually means a full mouth so I sketched one in. Although the skull was found in Australia, the coroner, a doctor who examines dead bodies, could tell by the dental work and thickness of bone that the man was probably from southern Yugoslavia. So I added Yugoslavian features.

"I also noticed that the bones at the back of the head were very large and thick which made me think that the man was probably very muscular. Although the man's identity hasn't been discovered, this method has identified many other people."

Now it's your turn. Try matching up the skulls and faces shown below. Look carefully at the shape of the forehead, jaw and chin.
Answers on page 96.

With the strong jaw at the base of its beak, a parrot can crack open its food.

The hooked beak of this hawk helps it pull apart its dinner if it's too big to be swallowed whole.

Mergansers use their toothy beaks to catch and hold onto their slippery dinner.

By jabbing its beak into soft mud, the curlew can reach prey buried too deep for other birds.

Who ate it?

To most scientists, the skull is an animal's most useful bone. It can reveal things such as feeding habits, relationships, intelligence and senses.

Put your detective skills to work in this puzzle. Can you match the bird skulls to the bird food shown? Think about which skulls look powerful, what food would need a great deal of force to eat, and what food would be hard to get at.
Answers on page 96.

How we use bones

Doctors are now working on ways to lengthen bones and help very short people or people whose limbs are different lengths. They've even found ways to use coral to help new bones grow. Bones can also be used to make china and glue. Some bones predict the future and there are some bones you can play with. You can even learn to draw better by… knowing your bones.

SPIRITS IN BONES

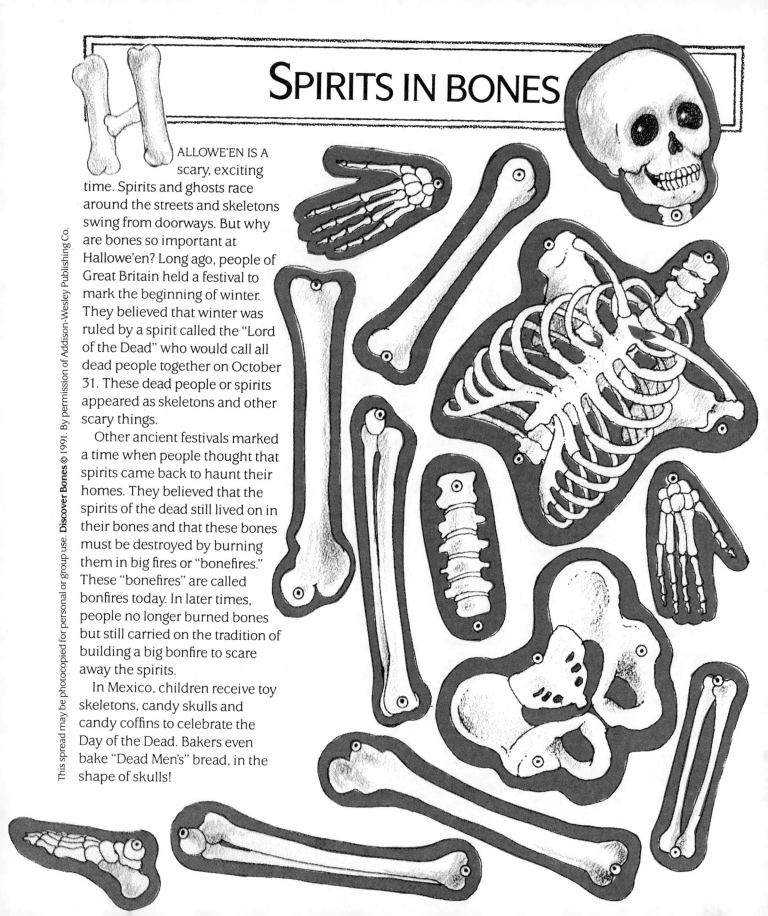

HALLOWE'EN IS A scary, exciting time. Spirits and ghosts race around the streets and skeletons swing from doorways. But why are bones so important at Hallowe'en? Long ago, people of Great Britain held a festival to mark the beginning of winter. They believed that winter was ruled by a spirit called the "Lord of the Dead" who would call all dead people together on October 31. These dead people or spirits appeared as skeletons and other scary things.

Other ancient festivals marked a time when people thought that spirits came back to haunt their homes. They believed that the spirits of the dead still lived on in their bones and that these bones must be destroyed by burning them in big fires or "bonefires." These "bonefires" are called bonfires today. In later times, people no longer burned bones but still carried on the tradition of building a big bonfire to scare away the spirits.

In Mexico, children receive toy skeletons, candy skulls and candy coffins to celebrate the Day of the Dead. Bakers even bake "Dead Men's" bread, in the shape of skulls!

Get ready for Hallowe'en

Make a skeleton to hang in your window.

You'll need:
○ *tracing paper*
○ *a pencil or thin marker*
○ *heavy white paper*
○ *scissors*
○ *glue, thread or paper fasteners*

1. Photocopy the bones shown here or trace them. To do that, place the tracing paper over each bone shown here and carefully trace each shape. Transfer your tracings to the white paper. (If your white paper is fairly thin, you might be able to trace the bone shapes directly onto it and avoid this tracing paper step.)
2. Cut out the bones.
3. Now join your model together. You can glue the bones together but if you join them with paper fasteners or thread, your skeleton will be able to move its arms and legs. Can you figure out where all the bones go? If you need some help, see page 10.

MEET THE BONE MAN

Up a creaky wooden staircase, past an enormous hollow-eyed elephant skull, and behind strings of dangling vertebrae, you'll find the bone man, Dr. Howard Savage.

Dr. Savage is a zoo-archaeologist, a scientist who works with animal bones. People from around the world send him bones and ask him to discover things about them, such as what animal the bone came from or how old the bone is. For instance, archaeologists working in the Arctic wanted to know the age of a whale rib they had found. They discovered the rib near a place where people had lived 1000 years ago and wondered if the bone was the same age. And why was it so strangely swollen?

When Dr. Savage X-rayed the bone, he discovered a bullet in it. He was able to identify the bullet as a type used early in this century. He knew the whale was still alive when the bullet struck it because the bone where the bullet was lodged had been irritated by it and had built up a swollen lump of bone around the metal. So Dr. Savage could tell the archaeologists that the bone was no older than the bullet, or about 80 years old.

And what's that hippopotamus doing in Dr. Savage's office? When Rosie the hippo died at the Metropolitan Toronto Zoo in Toronto, Ontario, Dr. Savage added her skeleton to his collection. All these bones help Dr. Savage and his students study ancient bones. If someone sends him a very old bone that might have belonged to a hippo or maybe a mammoth, Dr. Savage can look through his collection, match up the bones and identify the animal.

BONES ARE BEAUTIFUL

The Archer ~ Henry Moore

ARTISTS MAY BE INSPIRED BY A glowing sunset, the bold colours of flowers... or even bones! Henry Moore, a famous sculptor, was intrigued by natural forms such as pebbles and bones. You can see one of his sculptures above. Do you think it looks like a bone?

There are many amazing examples of bony art. Imagine carving a model ship out of bones left over from your dinner! French prisoners of war during Napoleon's time carved beautiful model ships from their dinner bones. So hunt around for "bone art" the next time you're at your museum.

The skeletal sculptor

All animals are different shapes because of their bony framework. It's really challenging to draw these different shapes so sometimes artists make clay models first to help them understand the shape of an animal better. Here's a recipe for making clay so you can make your own models, too.

Baker's clay

You'll need:

- 500 mL (2 cups) all-purpose flour
- 250 mL (1 cup) table salt
- a large bowl
- 175 mL (3/4 cup) water
- 30 mL (2 tbsp) glycerine (you can buy it at drugstores)
- food colouring

1. Mix the flour and salt together in the bowl.
2. Add the water and glycerine and mix.
3. Knead the mixture until it is smooth. Add the food colouring as you mix.
4. Make an animal or make bones, such as arm or hip bones. These are interesting works of art all by themselves.

5. Bake your sculpture at 150°C (300°F) until it is hard and light brown.

When it cools, you can paint your model with acrylic paints. To make it look shiny, coat it with a half-and-half mixture of white glue and water.

Painters and sculptors all know how important it is to understand body shapes. Artists long ago knew this too. Paul Toth, a museum artist, says, "Those artists couldn't have made such works of art if they didn't understand the anatomy (inner structure) of what they were drawing. To draw the outside, you have to understand the inside."

Mona Lisa ~ Leonardo da Vinci

Start by looking at your own body in a mirror. If you are "bony," you'll probably see your collar bone, ribs, spine, shoulder blades, hip bones and ankle bones quite clearly. Suck in your breath and look at your ribs. Then roll your shoulders forward to make your collar bone stick out. Try bending over to make your spine stand out. To sketch realistic positions, artists have to know how a body moves. Just how far do your bones, ligaments and muscles let you bend and twist?

Skeletal sketches

Put some of your observations to work.

You'll need:

○ a soft to medium pencil
○ a soft eraser
○ heavy white paper

1. Hold your pencil loosely and get to know what types of lines you can make with it—soft, hard, thick, thin.
2. Try sketching a part of your body, such as your hand. Feel your hand and you will notice that it's bony and straight on top and soft and rounded underneath. Use this information in your sketch.
3. Next time you're drawing at home or school, remember you can always use yourself as a model. Artists will often pose in a particular position themselves before they sketch it.

BONE UP

Leonardo da Vinci and Michelangelo, two great artists who lived hundreds of years ago, both studied anatomy. They probably knew more about the bones and muscles of animals and people than the doctors of their time. Today, artists with a special interest in anatomy sometimes become medical illustrators. One such artist, Dorothy Irwin, draws sports injuries. For instance, she had to draw the damage to the facial bones of baseball player Tony Fernandez when he was hit by a fast ball. Would she rather paint landscapes? "Not me. I think bones are awfully sculptural."

CLEANING BONES

IN A MUSEUM, SKELETONS ARE cleaned by "bugs." After most of the flesh is scraped off the skeletons, the bones are placed in the "Bug Room." There, thousands of tiny, furry beetle larvae feast until the bones are squeaky clean. But even beetles can be picky eaters. Sometimes scientists have to rub chicken bouillon on the meat to get the beetles to eat the really tough parts! Fortunately you can take the following shortcut.

You'll need:

- *heavy kitchen scissors, shears or small saw (get an adult's help)*
- *bones*
- *50 mL (1/4 cup) detergent*
- *a large pot filled with water*
- *a pipecleaner or skewer*
- *bleach (optional)*

1. Cut away as much meat as you can from the bones.
2. Cut off both ends of the bones. If you want to leave the ends on, get some help to drill holes at the ends before you cook the bones.

3. If you want to make small bone beads, cut up some of the bones at this point.

4. Rinse the bones until they are clean.
5. Add the detergent to the pot of water.
6. Gently simmer the bones for half an hour in the soapy water. Check the bones to see if you can pull off the rest of the meat. If not, boil the bones for another half hour.

7. Check the bones again. If they are clean, rinse them in cold water. If the bones still aren't clean, simmer them again in a pot of soapy water for half an hour and rinse. Be careful: too much boiling will weaken the bone.

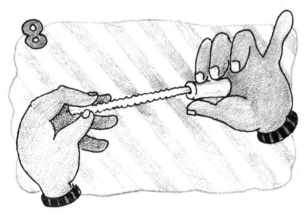

8. Push a pipecleaner through the bones to remove any marrow.
9. With an adult's help, add a capful of bleach to a pot of water. DO NOT SPLASH THE BLEACH OR GET IT ON YOUR HANDS. Bleach the bones for about 15 minutes. Rinse them well.
10. Dry the bones by leaving them uncovered at room temperature for two days, or bake at 90°C (200°F) for one hour.

Paint a bone

Painting on bones is one thing you can do with those gleaming white bones you've just cleaned. Nicki Nickerson from Lisle, Ontario, in Canada is famous around the world for her bone paintings. She got started when a sun-bleached bone she found one day reminded her of paper she'd been using for her watercolour paintings. Today she paints pigs, flowers, birds, cattle—just about anything—on bones she has prepared.

If you'd like to paint bones, start by cleaning some using the instructions here. Leave your clean bones outside in a sunny spot for a few weeks to dry and bleach them. Before you start painting, sand your bones until they are very smooth. When you've finished your painting, you may want to coat the bone with a protective covering such as varnish.

SKELETAL JEWELLERY

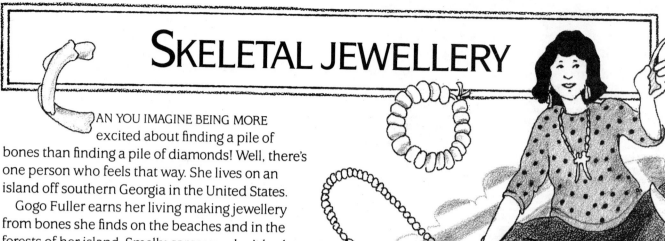

Can you imagine being more excited about finding a pile of bones than finding a pile of diamonds! Well, there's one person who feels that way. She lives on an island off southern Georgia in the United States.

Gogo Fuller earns her living making jewellery from bones she finds on the beaches and in the forests of her island. Smelly carcasses don't bother her at all. In fact, she feels that by using the bones, the animal's death isn't final. Gogo cleans the bones by putting them in a pen with hungry carrion beetles, or she hangs them off her dock for the crabs to pick clean.

You may not have shark vertebrae or snake ribs such as Gogo has, but you can make jewellery from other bones too.

Make a necklace

You'll need

○ sandpaper (wet and dry variety) or a file
○ clean bones (see instructions on pages 80-81)
○ cord or string
○ beads

1. Sand the ends of the bones until they are smooth.
2. Thread a 50 cm (20 inch) length of cord through the bones and beads in a design of your choice. To make a bracelet, you need about 25 cm (10 inches) of macrame cord.

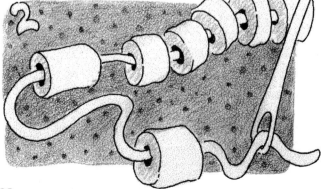

Here are some other examples of bone jewellery from around the world. Because bones come in so many different shapes and sizes, they have long been considered ideal material for jewellery. Even today, people wear bone jewellery.

Shape a bone

For native people, often the first step in making bone tools was to crack a bone open. A leg bone, for example, would be placed on a rock and struck in the middle with another rock, causing the bone to splinter. The sharp pieces were then shaped and polished with sandstone and used for many things such as needles, hairpins, awls, knives, chisels and combs. Other items such as spoons, scoops and hoes were made from the shoulder blades of large animals, such as deer and buffalo. Antler was used for spears, harpoons, clubs, spoons, bracelets and even purses. Because antler could be softened in water, it could be carved and decorated easily.

How easy is it to make a bone tool? Find out for yourself.

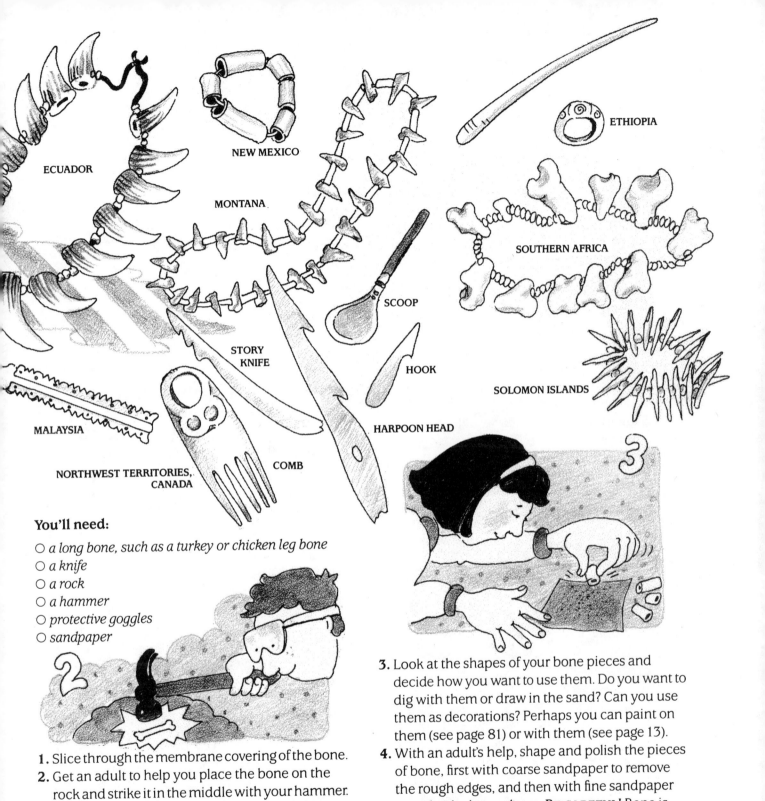

ECUADOR

NEW MEXICO

MONTANA

ETHIOPIA

SOUTHERN AFRICA

SCOOP

STORY KNIFE

HOOK

SOLOMON ISLANDS

MALAYSIA

HARPOON HEAD

NORTHWEST TERRITORIES, CANADA

COMB

You'll need:

○ a long bone, such as a turkey or chicken leg bone
○ a knife
○ a rock
○ a hammer
○ protective goggles
○ sandpaper

1. Slice through the membrane covering of the bone.
2. Get an adult to help you place the bone on the rock and strike it in the middle with your hammer. **BE SURE TO WEAR GOGGLES TO PROTECT YOUR EYES.**
3. Look at the shapes of your bone pieces and decide how you want to use them. Do you want to dig with them or draw in the sand? Can you use them as decorations? Perhaps you can paint on them (see page 81) or with them (see page 13).
4. With an adult's help, shape and polish the pieces of bone, first with coarse sandpaper to remove the rough edges, and then with fine sandpaper to make the bone gleam. **BE CAREFUL!** Bone is very sharp.

Medical miracles, sticky stuff and more

Sometimes elderly people who have broken their hips have to have them replaced with metal ones. Usually everything goes well, but occasionally their body starts to reject the metal hip and that can be very painful. Now artificial bone is coming to the rescue. The new material is so similar to real bone that it fools the bone into joining with it. This artificial bone is painted on the metal hip and within hours the artificial and real bones start to knit together. Doctors think that one day this artificial bone will be used to patch fractures and strengthen thinning bones.

What happens to those broken hip bones when they are removed and replaced? In one hospital, these bones were used to help a young boy whose spine was curving and putting pressure on some organs. The new bone was used to straighten his spine, giving him a chance for a normal life.

Another recent bony medical breakthrough is the use of coral to reconstruct a patient's bones. For instance, if a jaw bone has been destroyed by cancer, coral can be used to build up the jaw again. The best thing about coral is that the patient's body doesn't reject it, something that often happens with bone from other people. Soon after the heat-treated coral is implanted, the patient's bone enters the coral's maze of tunnels and fuses with it.

Bone marrow transplants are saving many children who have leukemia, cancer of the blood. Bone marrow from a donor is injected into the bones of the patient. If all goes well, the new bone marrow begins to produce healthy, cancer-free blood.

Bones are used to make some types of glue. First the bones are washed to remove all the grease. Next they are crushed, then cooked in a large tank until a gluey substance (the collagen) is released and becomes concentrated. This gluey liquid is heated over and over again until it becomes thick. Afterwards, unwanted particles are filtered out. The glue is then dried and sent to manufacturers.

Bone is an important ingredient in some expensive china. Ground-up clay and stone is mixed with a large helping of bone ash (made by burning animal bones) to make a paste. This thick, whitish paste is then shaped into chinaware (porcelain).

What's in the future for bones? Would you believe stretching and lengthening? Already doctors in Japan are working on it. They cut the bone, then hold the two cut ends in a fixed position with a small gap between them. Within 21 days, new bone grows to bridge the gap, making the bone a little longer. Over time, with this method, arms and legs can be stimulated to grow up to 25 cm (10 in.). This operation is especially important for people with very short limbs or limbs of different lengths.

Garden not doing so well? Squirrels munching your tulips? No problem...just sprinkle a little bone meal on the soil. Bone meal is made by drying bones and grinding them to a fine powder. The phosphorus it contains makes great fertilizer... and the squirrels can't stand it!

BONE UP

Bet you can't guess what's stored in Bone Banks! These banks contain a supply of bone that can be used to replace diseased bone, fix fractures and fill in any missing bone. The main source of this healthy bone is from people who have had a limb amputated.

Foretelling the Future

DO YOU EVER WISH YOU COULD TELL what will happen in the future? Maybe you've asked a Magic 8 ball questions such as: "What will I be when I grow up?" "Will I pass my math test?" "Does she like me?"

Thousands of years ago, Chinese people used bones to predict the future. They thought that animal bones had spirits in them that could tell them things, such as whether they would win a battle or recover from an illness.

How did they use these "oracle" bones to foretell the future? A person would ask a wise man or "shaman-priest" a question. The shaman priest would then singe a bone (or sometimes a tortoise shell) with fire to produce cracks in it. Cracks that turned upward could mean "yes" and cracks that turned down, "no," depending on what had been agreed upon at the start. It was a bit like playing "heads or tails," and deciding before you flip the coin what the meaning of a head or tail turning up will be.

Future basket

In some parts of Africa, the future was sometimes predicted with the use of a special divining basket filled with many objects. These might include a chicken bone, a shoulder bone of a turtle, a python vertebra, beads, small carved figures, etc. The diviner shook the contents of the basket and then looked at the patterns formed by the objects. Each object had its own meaning, depending on what objects it was lying next to. For example, if the python vertebra came up, it might mean "pain in the neck." Using the divining basket was a special religious activity.

HUNTING GUIDES

The Indians of Labrador and northern Quebec also burnt bones and then "read" the cracks and marks that appeared. However, they used the scapula (shoulder blades) of animals, such as caribou, to help them see into the future. Usually what they were trying to find out was where and when to hunt. The black spots, cracks and breaks that appeared on the burnt bone would often suggest rivers, mountains, lakes, trails and the location of different animals to the bone readers.

Take a look at this burnt shoulder blade. The bone reader could read it like a map:

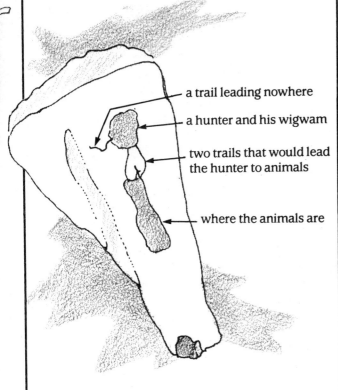

- a trail leading nowhere
- a hunter and his wigwam
- two trails that would lead the hunter to animals
- where the animals are

The hunters believed that animal spirits were guiding and revealing the future to them through this deeply religious practice of burning bones.

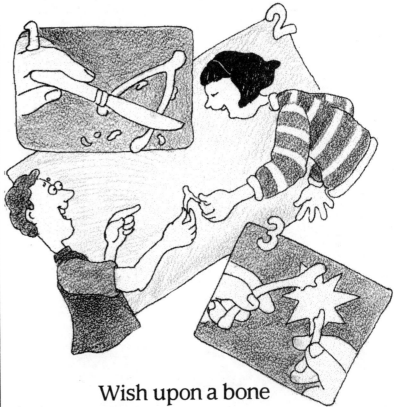

Wish upon a bone

Maybe you've tried to change your future a little bit by wishing on a bone—a wishbone. The practice is believed to have started in northern Italy about 2500 years ago. A chicken or turkey collar bone was dried in the sun and then a person would touch it and make a wish.

The next time you have a turkey or chicken for dinner, save the wishbone, then get ready to make a wish.

You'll need:

○ a sharp knife
○ a turkey or chicken wishbone
○ a friend

1. Carefully scrape the meat off the bone and let the bone dry out for a week or two.
2. Find a partner and both make a secret wish. Then with each of you holding one end of the bone as shown, pull the ends apart.
3. Whoever ends up with the bigger piece of bone, gets her wish!

BONE MYSTERIES

STORY HAS IT THAT THE INUIT PEOPLE are so skilled at making things out of bone, that once when a snowmobile broke down, a replacement part was carved out of bone!

The native people of North America have fashioned many beautiful and useful items out of bone.

Can you guess how the bone artifacts shown here were used? For some of the items, it may help you to think about cold, icy, snow-covered land.

1. To use this tool, the hunter first put the caribou anklebone in his mouth, then pressed down with it on the long piece of bone. A thong wrapped around this long bone and attached it to a bow-like object. Next the hunter pushed the "bow" back and forth, making the bone turn.

2. Inuit used this bone artifact mostly on bright sunny days. It's about 15 cm (6 inches) long and has two slits in the middle. Thongs made from animal skin held it in place. Are these clues too shady?

3. Can you guess what animal was hunted with these special tools? One tool probed the snow to find breathing holes in the ice. The second tool helped the Inuit hunters determine the size and shape of the hole. The third tool told the hunters when their prey had arrived at the hole.

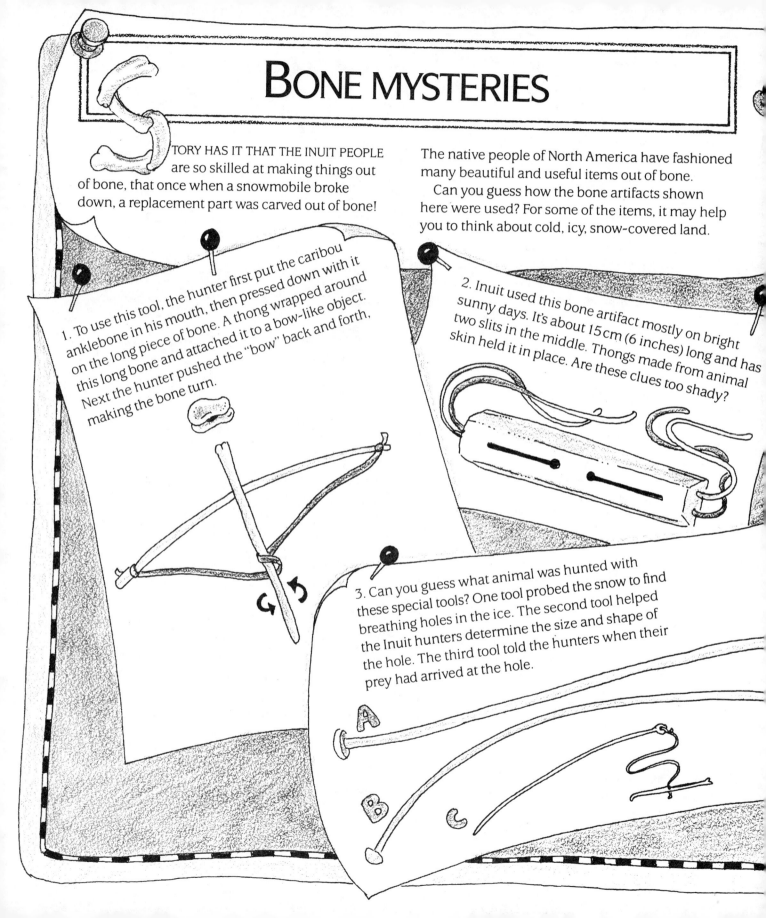

4. This item was made from a lynx jaw and was used by the Cree of northern Quebec. The long tooth was used to remove long lengths of worn-out rawhide thong that were part of something worn in the snow.

5. This tool was sometimes made from the hind leg bone of a moose or the arm bone of a bear. Native people of North America used it to prepare animal hides. Can you guess how?

Answers on page 96.

6. The Northern Plains Indians carved this item from the spongy part of buffalo leg bones. It is shaped to fit into the palm of a hand and was used in decorating hide clothing and teepees. Can you guess what it is?

OLD GAMES WITH A NEW LOOK

DID YOU KNOW THAT SOME OF THE games you play today were played by children hundreds, even thousands of years ago? Now a lot of your games are made of plastic or wood, but back then many games were made from bone.

Cup and pin

Native people of North America played this game and some also called it "toss and catch." A rabbit skull was used for the cup and a bone needle for the pin. The idea was to fling the rabbit skull up in the air and spear one of the holes in the skull with a pin or spike. Other native people made the game with caribou phalanges, fish vertebrae and even small animal pelvises.

You can make your own cup and pin game with lamb chop bones or a cardboard tube.

You'll need:

○ 4 or 5 lamb chop bones or a cardboard tube, about 13 cm (5 inches) long
○ 30 cm (12 inches) string
○ a pencil or stick

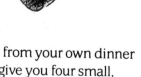

1. Collect lamb chop bones from your own dinner table or ask a butcher to give you four small, hollow bones.
2. Clean the bones as described on page 80. If you don't have lamb chop bones, cut the cardboard tube into pieces about 2.5 cm (1 inch) long and use these instead.

3. Tie one end of the string to a bone or tube and thread on the others.

4. Tie the other end of the string around the top of the pencil or stick. If the string isn't secure, dab a little rubber or plastic cement on the string where you have tied it. Let it dry.

5. Toss the "bones" and try to put the pencil (pin) through the hollow middles. See how many times you can do it, catching just one, then two or more at a time.

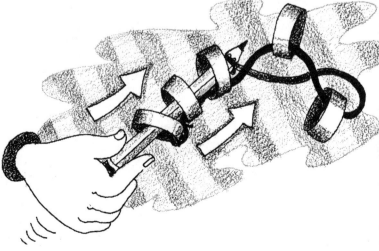

Spinning buzzer

Native people of the Arctic and Subarctic made this toy out of bone. They would thread sinew through a round bone and then spin it. As it spun, it made a great buzzing sound.

Here's how to make your own spinning buzzer.

You'll need:

○ scissors
○ cardboard
○ a piece of string about 2 m (6½ feet)
○ markers
○ a pencil

1. Cut out a cardboard circle with a diameter of about 7.5 cm (3 inches).
2. Punch two small holes about 1 cm (½ inch) on either side of the centre.
3. Poke a few small holes along the edge of the circle.
4. Decorate your disc with stripes, dots, swirls or any pattern you like.
5. Thread the string through the centre holes and tie the ends together.

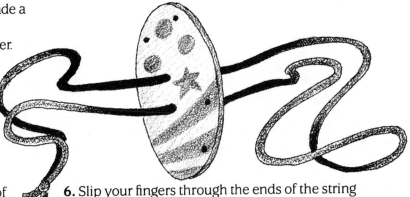

6. Slip your fingers through the ends of the string and wind up your disc by turning the string like a skipping rope.
7. Pull the string tight and watch the disc spin. Let the string relax while it spins the other way. Then pull again.

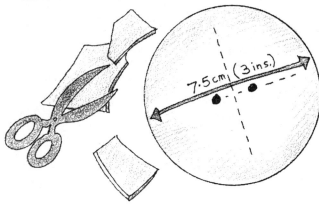

Glossary

Anatomy: The structure of an animal
Anthropologist: A scientist who studies people's customs
Archaeologist: A scientist who studies the people from the past by looking at the objects they left behind
Artifact: Any object made by humans
Calcium: An essential mineral for building strong bones and teeth
Carnivore: An animal that eats other animals
Cartilage: A smooth, flexible material found mainly on joint surfaces in the nose, ears and windpipe. In newborns, almost all bones are cartilage.
Collagen: A protein found in cartilage, tissue and bone
Compact bone: The hard, dense, outer layer of bone
Epiphyses: The growing ends of bones. Most bone growth takes place here.
Endoskeleton: An internal skeleton. Vertebrates (e.g., people) have an endoskeleton.
Exoskeleton: An external framework that provides support and protection for an animal's body
Fossil: Any trace of life from the distant past
Herbivore: An animal that eats grasses and other plant material
Invertebrate: An animal without a backbone
Joint: A place where bones meet
Ligaments: Short bands of strong, flexible tissue that connect bones together
Mammals: Vertebrates that produce milk and are covered with hair
Marrow: A soft fatty substance found in bones. Red marrow makes blood cells, while yellow marrow stores fat.
Midden: Ancient garbage dumps
Muscles: Tissues that can contract and expand to allow your body to move
Osteoporosis: A condition of brittle and fragile bones caused by lack of calcium and other bone forming minerals
Palaeontologist: A scientist who studies fossils and ancient forms of life
Skeleton: A strong, hard framework that supports a body
Spongy bone: The light but strong mesh-like bone that partially or entirely fills bones
Tendons: Cords of strong tissue that connect bones to muscles
Vertebrate: An animal with a backbone (vertebrae)

Activities and Experiments

ancestors' skeletons puzzle, 51
archaeological dig, 57
bake bones, 15
bend bones, 15
bird skull and food puzzle, 73
body benders, 29
bone-hard puzzle, 46-47
bone tools puzzles, 88-89
bony or boneless puzzle, 45
calcium recipes, 16-17
cleaning bones, 80-81
cookie excavation, 63
construct a skeleton, 76
dinosaur puzzle, 66-67
dog bones puzzle, 55
estimate your adult height, 23
finger flexing, 31
first aid, 39
foot impressions, 33
games and toys, 90-91
joints puzzle, 26
kite building, 42-43
model a giraffe, 11
owl pellet, 59
necklace, 82
paint a bone, 81
paint with bones, 13
prehistoric pottery, 53
saddle joint, 31
sculpting bones, 78
shape a bone, 83
skeleton identification, 69
sketching bones, 79
skull puzzle, 72-73
skull identification, 60-61
sleep and grow, 21
special bone puzzles, 70-71
struts, 35
tap your toes, 29
tendon tricks, 29
wishing with bones, 87

INDEX

age, how bones show, 68-69
anatomy, 79
ankle bones, 10, 20
ankles, 26-27
antlers, 46, 82
anvil bone, 35
ape, 51
arch of foot, 32, 33
archaeological digs, 56-59, 69
arm bones, 10-11, 27, 36-37, 40
armadillo, 47, 70
armoured animals, 10, 44, 47, 65, 70
art using bones, 13, 53, 78-79, 81, 82-83
arthritis, 27
artifacts, 57, 88-89
artificial bone, 84
artificial hand, 30
astronauts, 18

back pain, 24-25
backbones, 10-11, 21, 24-25, 26, 44, 71
baleen, 47
bats, 31
beaks, 60, 73
beams, 34
beavers, 61
beetles, 45
biceps, 29
birds, 40-41, 43, 58, 60, 70-71, 73
blood vessels, 12, 46
Bone Banks, 85
bone china, 85
bone marrow transplants, 85

bone meal, 85
boneless animals, 44-45
bonfires, 76
bound feet, 33
bow legs, 16
boys and girls,
    adult height and weight, 23
    differences in bones, 20, 68-69
breastbones, 11, 13, 20, 41
brittleness of bones, 15, 19
broken bones, 36-39, 67, 84, 85
Bug Room, 80
buried bones, 55, 56-59, 69

calcium, 14, 16-19, 36, 46, 62
calcium phosphate, 14, 46
Canadarm, 27
caribou bones, 87, 88
carnivores (meat eaters), 55, 61, 66-67
carpals, 69
cartilage, 14, 20, 21, 27, 69
carving, 52, 78
casts, 36-37
cheekbones, 72
chinaware, 85
chiropractors, 24
chitin, 44, 45
clavicles, 11
cleaning bones, 80-81
coccyx, 11
cochlea, 35
Coelophysis, 65
collagen, 14, 15, 46, 85
collar bones, 11, 20, 87

compact bone, 12, 34
components of bones, *see* ingredients
Compsognathus, 64
coral, 47, 84
crabs, 45
crocodiles, 10
curlews, 73

deers, 61
dinosaurs, 27, 64-67, 70
discs (spinal), 21, 24
dogs, 27, 54-55, 58

ear bones, 35
elephant fish, 70
elbows, 20, 26-27, 38
epiphyses, 20
estimating adult height
    and weight, 23
Euoplocephalus, 65
evolution, 50-51
exercises, 18-19, 25
exoskeletons, 44-45
eye sockets, 72

facial features, 72-73
fat in bones, 13
feet, 10, 20, 32-33
female skeletons, 68-69
femur, 10
fibula, 10
finger bones, 31, 40
first aid for broken bones, 39
fish, 70
flat feet, 33

93

flexibility of bones, 15
flying machines, 40
foods with calcium, 16-17
foot bones, 10, 20, 32-33
fossils, 58, 62-63, 64-65
fractures, 36-39
frogs, 28, 61
Fuller, Gogo, 82
funny bone, 38
fusion, 20-21, 40, 68-69

games and toys, 90-91
garbage heaps, 56, 57
gerbils, 61
glue, 85
growth of bones, 20-23, 46, 68-69, 85
growth hormone, 22

Hadrosaurs, 65
Hallowe'en, 76
hammer bone, 35
hands, 10, 27, 30-31
hares, 28, 29
hawks, 73
healing broken bones, 36-38, 46, 67, 84, 85
height, 22-23
herbivores (plant eaters), 61, 67
Herculaneum, 69
hermit crabs, 45
herons, 43
hip bones, 20, 84
hip replacement, 84
hollow bones, 40-41, 71, 89
Homo australopithecus, 51
Homo sapiens, 51
horns, 46
humerus, 11, 40
hunting, 87, 88

identifying bones, 10-11, 58, 59, 60-61
ingredients of bones, 13, 14-16, 36, 46, 85
insects, 44-45
invertebrates, 44-45, 62
Iroquois culture, 52
ivory, 46

jaws, 26, 55, 71, 72
jewellery made of bones, 82-83
joints, 20, 26-27, 30-31, 37, 71, 84

keelbone, 43
kites, 41-43
kneecap, 10
knees, 20, 26-27
knuckles, 27

leg bones, 10, 28, 43, 66
ligaments, 27, 33, 71
loons, 71

male skeletons, 68-69
mammals, 54, 61
marrow, 13, 85
meat eaters (carnivores), 55, 61, 66-67
medial arch, 32, 33
medical illustrators, 79
mergansers, 73
metacarpals, 10
middens, 56, 57
Moore, Henry, 78
mouse kits, 59
mouse skulls, 58
mummies, 61, 69
muscles, 29, 31, 37, 44, 66, 67
muskrats, 61

Neanderthals, 50-51
nerves in bones, 38
Nickerson, Nicki, 81
nose cavity, 72-73

oracle bones, 86
osteoblast cells, 36
osteoclast cells, 36
osteoporosis, 19
owl pellets, 59

Pachycephalosaurs, 70
pacu fish, 70
painting with bone, 13
painting on bone, 81
parrots, 73

patella, 10
pelvis, 11, 13, 28, 50, 69
penguins, 43
phalanges, 10
phosphorus, 14, 15, 85
pincer grip, 30
plant eaters (herbivores), 61, 67
plates of armour, 10, 47, 70
porcelain, 85
porpoises, 31
posture, 24-25, 50
pottery designs, 53
predicting the future, 86-87

radius, 10, 27
replacing bones, 84, 85
ribs, 11, 13
rickets, 16

saddle joint, 30-31
Savage, Howard, 77
scallops, 45
scapula, 11, 87
sculpture, 78-79
Seismosaurus, 64
shaft of bones, 34
shells of turtles, 10, 28, 29, 47
shin bone, 10
shoes, 33
shortest person, 22
shoulder blades, 87
silica, 62
skeletons, 11, 20, 28, 40-41, 51, 54-55, 59, 66-67, 68, 76-77, 78-79
sketching bones, 79
skulls,
  animal, 55, 58, 60-61, 70, 73
  human, 11, 21, 26-27, 35, 50-51, 70, 72-73
skyscrapers, 34
snakes, 61, 71
soft bone, 36
spinal cord, 10, 11, 24
spines, 10-11, 21, 24, 40, 47, 55, 71
spongy bone, 13
squid, 45
squirrel skulls, 58
sternum, 11, 41

stingray, 47
stirrup bone, 35
strength of bones, 13, 14, 18-19, 40, 46
structure of bones, 12-13, 34-35, 36, 38, 40-41, 71
struts, 34, 35, 40
sunlight, 16
sutures, 27
swans, 43
synovial fluid, 27

tail bone, 10, 11, 55
tails, 41, 55
tallest person, 22
teeth, 46, 55, 61, 67
tendons, 20, 27, 29, 31
thickness of bones, 69

thumbs, 30-31
tibia, 10
tissue on surface of bone, 12
tools from bones, 13, 52-53, 58, 82-83, 88-89
tortoises, 10
toys, 90-91
triangles, 34-35
turtles, 28, 29, 45, 47, 60
tusks, 46

ulna, 10, 27
ulnar nerve, 38
Ultrasaurus, 64

vertebrae, 10, 11, 13, 21, 24, 26, 55, 71

vertebrates, 21, 62
vitamin D, 16

walking, 32
water in bones, 15
weight of adults, 23
weight of bones, 13, 23, 69, 71
weightlessness, 18
whale bone, 47
wings, 40-41, 43
wishbones, 87
wolves, 61
woodpeckers, 70
worms, 45
wrist bones, 68-69
wrists, 26-27

ANSWERS

Treat your back right, page 25

Dog A is letting other parts of his body help support his back. While sleeping on your back is okay, Dog B's head is too high.

Joints...where bones meet, page 26

1. Your elbow and knee joints are like hinges.
2. Your shoulder and hip are like ball and socket joints.
3. The top two vertebrae in your neck are like a pivot joint. The second vertebra, the axis, has a peg in the middle that fits through the first vertebra, the atlas. You could say that your head is bolted on with this nut and bolt arrangement.

Bony or boneless, page 45

1. A beetle has an exoskeleton and no bones.
2. Instead of bones, the crab has an exoskeleton with many joints. After it has shed its old exoskeleton, it must hide for a few hours until the new exoskeleton hardens.
3. The turtle has an inner skeleton of bones and an outer bony exoskeleton too.
4. The hermit crab has to live in an old mollusk shell to protect its soft body since it has no bones.

A bone-hard puzzle, page 46

1. They are bone.
2. Horns aren't bone. They are made of a hard material called keratin, the same material your fingernails contain.
3. Teeth are made of dentin, a substance related to bone but not bone.
4. The armadillo's body is covered with bony plates.
5. The turtle's shell is made of bone.
6. The spine of the stingray is bone.
7. Millions of tiny corals live together and build a hard protective covering of limestone, using minerals in the water. Coral isn't bone.
8. Whale bone isn't bone, it's a horny elastic material that hangs from the whale's upper jaw and is used to strain the tiny sea creatures they eat.

Digging up our past, page 51

The changes in the size and shape of the skull indicate an increasing brain capacity from ape to *Homo sapiens*.

H. sapiens' widely-spaced eye sockets allow us to have a broader range of vision—we can look to the front and the sides without turning our heads. A larger range of vision makes it easier to find food, spot enemies, etc.

The jaw bone and teeth change in size and shape as the diet changes. For example, apes eat mainly plants and insects but *Homo australopithecus* probably ate more meat. *H. sapiens* eat a variety of foods but our chewing capacity doesn't need to be as great as that of our ancestors' since our food is usually cooked and is easier to eat. As well, we don't need a strong jaw since we use weapons for defense, not our teeth, as apes do.

The size and shape of the hip bones has changed to allow *H. sapiens* to walk upright. In apes, you can also see that the arms and legs are almost exactly the same length, which makes walking on all fours easy. Our legs are longer and stronger than our arms because we use our legs for walking.

Humans are also becoming taller thanks to better nutrition.

Dogs and their bones, page 55

The greyhound is the slender racer and the basset hound is the keen-nosed hunter.

Build a dino, page 66

Look for broken ribs and a broken tail bone. *Diplodocus* was a plant eater (herbivore).

Only your archaeologist knows for sure, page 68

The X-ray on the left belongs to a two-year-old, the X-ray on the right to an adult (one carpal bone is hidden behind the others).

Boy or girl, page 69

The skeleton on the right is female. If you look closely at the pelvis, you'll see that it's broader than the male's pelvis. That's to make childbirth easier if the woman has children. In females, the pelvis is usually wider than the width across the shoulders. A man usually has wider shoulders than pelvis. The bones of males also tend to be thicker and heavier than female bones.

Special bones for special jobs, pages 70 and 71

The fish on the left is a pacu fish and has strong teeth to help grind its food. The elephant fish, on the right, can reach food in crevices.

Your bottom jaw does most of the work.

You look just like...your bones, page 72

1.d, 2.e, 3.b, 4.c, 5.a.

Who ate it, page 73

1.d, 2.c, 3.a, 4.b.

Bone mysteries, page 88

1. A drill: This drill could make a hole in any kind of bone or wood. The end of the drill was a sharp bone or iron point.
2. Snowgoggles: The sun reflecting off the white, Arctic landscape can be very hard on eyes.
3. These devices helped the Inuit hunt seals. The first artifact was used to find a seal's breathing hole. Then the breathing-hole searcher was turned around in the hole to get an idea of the hole's shape. This helped the hunter figure out on which side of the hole the seal was likely to come up. The third tool had two parts. One was placed in the breathing hole and the other was anchored in a snowbank. When the seal came up for air, it caused the rod in the snow to move, letting the hunter know that the seal was there. The hunter then quickly harpooned the seal.
4. The lynx jaw was used for removing worn webbing from snowshoes.
5. The scraper was used to scrape flesh and hair off animal skins so they could be made into clothing, rope or moccasins.
6. It's a painting tool. One tool was used for each colour.